Population Systems

Alan A. Berryman • Pavel Kindlmann

Population Systems

A General Introduction

 Springer

Alan A. Berryman
Washington State University
Pullman
USA

Pavel Kindlmann
Biodiversity Research Centre
Institute of Systems Biology
  and Ecology ASCR
České Budějovice
Czech Republic

Additional material to this book can be downloaded from http://extras.springer.com.

ISBN 978-94-007-9836-6   ISBN 978-1-4020-6819-5 (eBook)

Printed on acid-free paper

9 8 7 6 5 4 3 2 1

springer.com

Dedication

To
Rachael, Ashley, Annie, Lucka and Petr
Our delightful contributions to the population problem
and in memory of
Thomas Malthus
who saw the ultimate consequences

Preface to the First Edition

I had taught courses in applied ecology, population dynamics, and population management for many years and, like many of my colleagues, had grown accustomed to the blank stares of my students as we wove our way through the confused semantics and intricate concepts of traditional ecology and wrestled with elaborate mathematical arguments. I searched in vain for a central unifying concept on which to organize a theory of population ecology until, 30 years ago, I read a small book of essays edited by John Milsum of McGill University entitled *Positive Feedback – A General Systems Approach to Positive/Negative Feedback and Mutual Causality*. Stimulated by the articles in this book, particularly those written by Milsum, M. Maruyama, and A. Rapoport, I began to structure my lectures around the central ideas of general systems theory. I first used this approach in my graduate courses in population dynamics and population management and then, encouraged by the results, in my undergraduate course in forest entomology and to teach population dynamics to practicing foresters. Almost without exception, my students found the general systems approach intuitively reasonable and easier to understand than traditional teaching methods. Even undergraduates seem to grasp the fundamental principles quite rapidly and, more important, to realize that a general understanding of population systems is an essential part of their education. These reactions by my students, and their continued encouragement, led me to write this book.

This book is concerned with the general principles and theories of population ecology. I have attempted to derive these from a basic understanding of how general systems behave together with observations of the behavior of real population systems. Unlike some of my colleagues, I am convinced that the rules governing the dynamics of populations are relatively simple, and that the rich behavior we observe in nature is a consequence of the structure of the system rather than of the complexity of the underlying rules. This is aptly demonstrated by the "Game of Life" discussed in Chapter 1. In this chapter I have tried to provide a basic framework for analyzing the structure and dynamics of systems in general, using a simplified interpretation of general systems theory. From this perspective we then examine the dynamic behavior of single-species populations in Chapter 2 and develop an elementary feedback model of the population system. In Chapter 3 this single-species model is refined and generalized by examining the mechanisms of population regulation, and graphical procedures are developed for evaluating the behavior of populations inhabiting variable environments. These graphical methods are then applied to the analysis of interactions between two species, including mutualistic, competitive, and predator-prey systems, in Chapter 4. Then, in Chapter 5, we extend our dimensions to examine spatial effects on population behavior, and in Chapter 6 we take a brief look at communities composed of many interacting species.

Because I am convinced that all of us in this overcrowded world should be familiar with the basic concepts of population dynamics, I have attempted to write this book in a way that is comprehensible to the undergraduate student and layman, as well as being stimulating to the graduate student, professional population manager, and teacher. For this reason I have tried to avoid much of the ecological jargon and the complicated mathematics, which abound in the literature. The mathematics I have used is mostly elementary algebra, though more complicated arguments are presented, for those who wish to delve more deeply, in notes at the end of each chapter.

Although this book is of a theoretical nature, it is written with the applied ecologist and population manager in mind. At heart I am an applied ecologist, but I am also convinced that a firm theoretical background is essential if we are to make sound decisions concerning the management of our renewable resources and to anticipate the subtle consequences of these decisions. Managers frequently have to deal with population systems that are undefined, or only partly defined, by empirical data. Under these conditions they must rely on an intuitive understanding of the processes and interactions of the system. Population theory forms a basic framework on which this understanding can be built with the help of experience and an inquiring mind. This is not to say that a detailed knowledge of the properties and behavior of specific population systems, as well as the tactical tools available to the manager, are not equally important to the applied ecologist. Ideally this book should be used as a supplement to a specific text in courses aimed at the management of forest, range, wildlife, fish, or pest populations.

The theme throughout this book is populations interacting with their environments, and its main message is that populations of plants and animals can be intelligently managed if the general rules governing their behavior are clearly understood. If there is some urgency in my message it is because of my concern for this overcrowded planet and for our threatened renewable resources. Should this book contribute to our understanding of the immense problems we face, my time will have been well spent.

A. A. B.
Pullman
Washington
February 1980

Preface to the Second Edition

In the early 1980's, when I was at the beginning of my carrier of a theoretical ecologist, I came across a blue book called Population Systems. The intuitive approach adopted here was clearly distinct from all other books on mathematical modeling of population dynamics available at that time. Instead of masses of equations, followed by calculation of equilibria and their stability, the topic was explained here using drawings of isoclines and reproduction planes and the reader was asked to use visualization (and sometimes even something like intuition) to predict the behavior of complex biological systems. Despite my previous training in mathematics, I was amazed by the amount of practical interpretations, which could be derived from the models by means of this purely "visual" approach. I began to understand that mathematicians, by using explicit forms of their equations, often indulge themselves in complicated calculations, which then obscure the biologically interesting predictions of their models.

I soon found I was not alone. Many of my colleagues oriented in theoretical ecology, which had been trained as biologists (including Tony Dixon, Vojta Jarošík and many other people mentioned below in the Acknowledgements), found this inconspicuous book very appealing for exactly the same reason – intuitive approach to the problem. The book, however, did remain alone for more than 25 years. At least, I am not aware of any other book using the reproduction plane approach to such an extent, as done in Population Systems. Thus I was not surprised, when Alan Berryman was invited to publish its second edition. And I was very much honored and excited, when he agreed to accept me as a co-author, who would contribute negligibly by helping him with the revision.

Thanks to the unique "reproduction plane" approach, the main text did not require any dramatic changes, as most of it still stands – even more than 25 years after is has first seen the light of the world! Admittedly, some expressions, like "if you have a programmable pocket calculator available", became rather obsolete. We decided to accompany the book with a CD, where the reader can find lots of useful EXCEL files, illustrating the statements made in the main text and showing some examples of continuous systems. We refer to this disk, whenever appropriate. The introductory file appears automatically after the CD has been put into the drive – and the student is then instructed about how to use the other files. We also added a few new references and examples, which were published since the first edition, but are aware that we certainly did not include all those worth citing.

We hope that this slightly updated version of the classic book might find its place in the fast-growing array of literature on mathematical ecology.

P. K.
České Budějovice
August 2007

Acknowledgments

This book represents a synthesis of information and ideas obtained from many different sources, which have been blended with the particular (peculiar?) views of the senior author. The origins of many of these ideas have long been lost, but they include the contributions of well-known and unheralded ecologists, mathematicians, and systems scientists. The early thinking of the senior author was greatly influenced by his teachers, first at Cornwall Technical College, where Gordon Ince guided his birth as a biologist, and then at Imperial College of Science and Technology, London, and the University of California, Berkeley. At these latter schools his fascination with population ecology flourished under the tutelage of O. W. Richards, T. R. E. Southwood, N. Waloff, R. W. Stark, C. B. Huffaker, and D. W. Muelder. His interest in ecology developed during the Great Debate between A. J. Nicholson and H. G. Andrewartha, and their adherents, and it has been sustained and enriched by the contributions of C. S. Holling, R. M. May, R. Levins, and many, many others. As mentioned in the preface to the first edition, the conversion of the senior author to a general systems approach was brought about by reading the delightful book edited by J. H. Milsum, but his friend and colleague L. V. Pienaar also played an invaluable role in his education.

Many of the ideas presented in this book were forged by years of debate and argument with friends and colleagues. These sometimes vigorous personal interactions have provided the feedback which has nourished the thinking of the senior author and include discussions with A. S. Isaev, R. M. Peterman, G. E. Long, A. P. Gutierrez, K. J. Stoszek, D. L. Dahlsten, R. R. Sluss, E. C. Zeeman, J. A. Meyer, L. R. Ginzburg, M. P. Hassell, W. Baltensweiler, P. Carle, N. C. Stenseth, J. A. Logan, D. L. Wollkind and many others too numerous to mention. Hopefully they will not feel slighted by the failure to mention them by name. The graduate students of the first author have also contributed much to his thinking as their fresh young minds challenged conventional wisdom. He has taught them little but they have learned much together.

The junior author would like to mention two people at the first place: M. Rejmánek, who was the first person that introduced him to the concepts of mathematical biology, and A.F.G. Dixon, his lifetime friend and collaborator, who initiated his interest in modelling the life history strategies and whose ideas greatly influenced his further development as a scientist. The thinking of the junior author has also much profited from interesting and fruitful discussions with many colleagues, in the first place with (alphabetically): R. Arditi, K. Basnet, J. Baudry, F. Burel, J. Frouz, L. J. Gross, R. Harrington, J.-L. Hemptinne, M. Hulle, J. Lepš, A. Mackenzie, V. Novotný, J. S. Pierre, M. Plantegenest, R. B. Primack, D. Roberts, R. Tremblay, S. A. Ward, W. W. Weiser, D. F. Whigham, J. H. Willems and H. Yasuda. Personal interactions with other colleagues and students have provided him lots of intellectual stimuli and include discussions with (alphabetically): S. Aviron, Z. Balounová, S. Bečvář, B. Bhattarai, J. Blízek, C. N. Brough, P. Ceryngier, P. Cudlín, C. A. Dedryver, I. Dostálková, M. Grycz, F. Halkett, J. Havelka, I. Hodek, J. Holman, K. Houdková, V. Jarošík, P. Janečková, Y. Kajita, R. Kundu, Z. Mráček, O. Nedvěd, M. Okrouhlá, P. K. Paudel, J. Rajchard, P. Řezáč, A. Rico, Z. Růžička, S. Sato, R. Sequeira, P. Šmilauer, M. Špinka, K. Spitzer, B. Stadler, M. Stříteský, I. Stuchlíková, K. Wotavová, and many others.

They have all contributed, but three students of the senior author, G. C. Brown, K. F. Raffa, and R. H. Miller, and three students/colleagues of the junior author, O. Ameixa, J. Jersáková and I. Schodelbauerová, have given most because they read the book and made useful suggestions for its improvement. The book was also read by H. W. Li, R. W. Stark, A. P. Gutierrez, and D. R. Satterlund, and their constructive criticism and thoughtful suggestions have been of great help in preparing this manuscript. The book has also benefited from comments by students taking the graduate class in Population Management, and by professional foresters taking the short course in Population Dynamics in the U.S. Forest Service Continuing Education Program, both by the senior author, and particularly from the detailed review of P. J. Castrovillo.

Finally, we would like to thank those who kindly allowed us to use their original illustrations; W. C. Clark, D. D. Jones, and C. S. Holling and Plenum Press for Figure 5.4, C. B. Huffaker for Figure 5.8, and G. E. Long and Elsevier Scientific Publishing Company for Figure 5.12. All the other figures from published material were redrawn by J. Singleton and are acknowledged in their captions. We thank the authors for their permission to redraw their figures. The quotation from J. M. Keynes, which ends this book, was printed with the kind permission of Granada Publishing Ltd. The second edition was supported by the grant No. LC06073 of the MSMT.

A. A. B. and P. K.

Contents

Part I
Population Systems

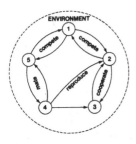

Populations are made up of individual organisms, which interact and communicate with each other as they pursue their normal lives. For example, individuals mate, compete for scarce resources, and cooperate to capture prey or escape being eaten. As a result of these interactions individuals reproduce, move and die and these processes cause the population as a whole to behave in certain ways - populations grow, decline, or remain steady.

Any system you wish to consider, a television set or an automobile, is basically composed of a set of interacting parts that together produce patterns of behavior, which are characteristic of the system. This behavior is determined by the rules of interaction, and the overall structure of the interaction network. Populations, therefore, can be thought of as particular kinds of systems with their own rules and structure, which – nevertheless - obey certain general system laws.

In the first chapter of this book we will take a brief excursion into the theory of dynamic systems in order to understand the properties and behavior of systems in general. Then these concepts will be applied to the analysis of single-species populations. In Chapter 2 a very simple model is developed from observations of the behavior of natural populations, which will help us to understand the fundamental rules of population growth. Then in Chapter 3 a more detailed model of the population system is created, along with a methodology for analyzing the behavior of dynamic population systems inhabiting variable environments.

Chapter 1
A Brief Look at Systems in General

The theory of dynamical systems originated in the engineering sciences as a way of describing and designing complex mechanical and electronic systems. It has since found increasing use by military, economic, and industrial strategists, as well as biologists, as a way of gaining insight into the structure and function of complex systems. In this first chapter we outline some of the elementary concepts and principles of dynamic systems theory as a prelude to our investigation of population systems. We have tried to avoid engineering jargon as much as possible and have freely modified some of the more rigorous concepts to suit the particular needs of population ecology, hopefully without losing the original intent. Our aim is to use the theory to gain a better understanding of population ecology and management and, thus, we have glossed over or ignored much of the formality and detail (references to more technical treatments are given in Note 1.1 at the end of this chapter).

1.1. What is a System?

A system is an assemblage of objects or components which interact, intercommunicate, or are dependent on each other so as to operate as an integrated whole. For example, the human body is a system composed of many interacting and interdependent organs, as is a television set made up of electronic parts and an automobile with its mechanical and electrical components. Now you may have realized that these systems are themselves composed of a number of discrete subsystems – your body has a nervous system, a circulatory system, a digestive system, and your car has a fuel system, an ignition system, and so on. The definition of a particular system, therefore, depends as much on the interest and perspective of the individual observer as on any intrinsic property of the thing being observed. The system exists in the eye of the beholder, so that the TV set is *the* system to the repairman while the TV network is the executive's system. The cell is to the microbiologist what the organ is to the physiologist, the organism to the behaviorist, and the population to the ecologist.

Although we can view things with different degrees of fineness, or resolution, no view is completely independent of the other. The organ can no more function

A.A. Berryman, P. Kindlmann, *Population Systems: A General Introduction*
© Springer Science+Business Media B.V. 2008

without its organism than can an automobile without its ignition system or a TV set without an electrical system. Thus, most systems are, in truth, only parts of larger systems, which are themselves parts of larger systems, and so on *ad infinitum* (Figure 1.1).

We can approach the problem of resolution by allowing ourselves the freedom to define a system according to our particular interest, and to treat the larger universe, of which our system is part, as an external *environment*. This environment supplies all the materials, energy, and information needed to make the system work. Hence, the human body is supplied with food, oxygen, water, shelter, and contact with other humans by its environment. Similarly, the TV set runs on its external source of electricity and radio waves and the automobile on gasoline, oil, water, and oxygen. All these *resources* in the environment are considered to be *inputs* into the system. Inputs may vary with time (then they are called variables) or remain invariant with time (then they are called constants), but whatever is the case, they control or activate the components of the system and enable it to function.

Environmental inputs may sometimes disrupt or even destroy the system. Most mechanical and biological systems have certain design tolerances, which cannot be exceeded without seriously affecting their operation. Overloads of otherwise essential resources may have disastrous effects - too much electricity blows the television set, too much gasoline floods the carburetor, too much water drowns the

Fig. 1.1 A hierarchy of systems

animal - and catastrophic events in the environment, such as earthquakes, hurricanes, and volcanic eruptions, can seriously disrupt or destroy the natural ecosystem. In other words, there are certain environmental inputs, which are usually very rare, that the system is not designed to deal with. When these rare events occur the system can be seriously disrupted or even destroyed.

Systems may also contribute materials, energy, or information to their environments. For instance, humans exude feces, urine, carbon dioxide, heat, and knowledge, while their automobiles emit sulfur dioxide, carbon monoxide, and other gases. These contributions from a system to its environment are called, reasonably enough, *outputs*. On occasion outputs can have serious effects on the environment, which may even threaten the system that produced them. For example, waste products from humans and their agricultural, industrial, and transportation systems pollute the environment and, in large quantities, may make it unfit for human existence.

Our basic ideas concerning a system and its environment are summarized in Figure 1.2. In this diagram we have separated the system from its environment for reasons of clarity. In reality, of course, the system operates within its environment. We consider the subject system to be composed of a set of interacting or interdependent parts, which are delineated by a boundary defining that particular system. The larger universal system (or systems) within which the subject system exists is defined as the environment. Inputs into the system from its environment supply the materials, energy, and information needed to make it run or which may disrupt or destroy it. The system may produce its own materials, energy, or information outputs, which flow back into the environment and may feed back to affect the subject population itself.

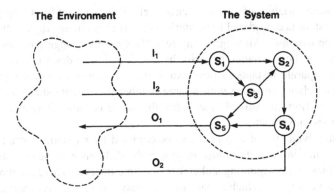

Fig. 1.2 A system composed of five components, S's, two of which are affected by inputs from the environment, I's, and two of which produce outputs into the environment, O's

1.2. The State of a System

At a particular instant in time a system can be viewed as a static assemblage of parts, much as a photograph is a static representation of a moving object. The system, with all animation suspended in space and time, can be described accurately because all the moving parts are frozen in place. Such a description characterizing a system at a given instant in time is called a *state description*.

Although the state of a system is described by the condition of its component parts, some of them may not change appreciably over time and, hence, are not particularly interesting from a dynamic point of view. For instance, describing a person as having a head, torso, arms, legs, etc., is not very meaningful because most of us have them, and their general nature does not change much from time to time or place to place. However, describing a person as being in love, in poor health, or in a hurry is of more interest because these conditions can vary considerably. Thus, we commonly use expressions such as "state of mind" and "state of shock" to characterize particular conditions that may change drastically in the next moment. Components of the system that change in time are called variables, and those that we use to characterize the state of the system are known as *state variables*.

Most complex systems possess hundreds or even thousands of state variables and it is usually impractical, or even impossible, to describe the condition of them all. The art of diagnosis, then, is deciding which of the state variables should be used to describe the state of a particular system. For example, the general state of your health can be characterized by measuring your blood pressure and by analyzing a sample of blood and urine. Thus, state variables are usually chosen because they are the most sensitive indicators of the changes that interest the analyst.

1.3. Dynamical Systems

When the state variables of a system remain relatively constant for a long period of time, the system is considered to be *static*, while if they change rapidly the system is said to be *dynamic*. Although many real-life systems change continuously, we often represent their dynamic behavior by a series of state descriptions made at a number of separate instants in time. An analogy is a movie, which represents a continuously changing scene with a large number of separate static photographs taken at very short intervals of time. When the movie is shown it gives the *illusion* of continuous movement.

Since the dynamics of a system can be depicted by its static portrait taken at discrete instants in time, its *change in state* is the difference in the condition of its state variables at the beginning and end of one time interval. There are three possible qualitative ways in which a state variable may change: It may increase (+), it may decrease (−), or it may remain unchanged (0). The way in which a particular variable changes, and the magnitude of the change, is determined by its interaction

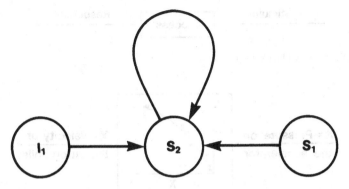

Fig. 1.3 Interactions that may affect a change in the state of a variable; the state variable S_2 is influenced by another state variable, S_1, by an environmental input, I_1, and by itself

with other state variables, with inputs from the environment, or with itself (Figure 1.3). Therefore, interactions between state variables and inputs control the dynamic behavior of the system.

1.4. System Diagrams

There are two basic conventions for representing the relationships between the variables of a system: *flow graphs* and *block diagrams*. In the former, variables are represented as circles, or nodes, and the flow of matter, energy, or information between them by arrows (Figure 1.3). Flow graphs are particularly useful when the flows are simple linear functions of their variables, and when we are dealing with systems where the variables are in equilibrium; that is, they remain more or less constant with time. We will use flow graphs in only one chapter of this book, when considering communities of organisms that are near to equilibrium (Chapter 6). In the remainder of the book we will use the block diagram convention because it is generally more flexible and easier to apply to population systems.

The basic components of block diagrams are boxes, which represent processes or mechanisms, and arrows, which represent the variables that operate the processes (Figure 1.4). Variables that enter a box stimulate the process, which gives rise to a response in the form of a variable leaving the box. Thus, arrows entering boxes represent *stimulus variables*, whilst those leaving boxes represent *response variables*. These terms are used whether the variables are state, input, or output variables.

We can view the processes or mechanisms of the system as subsystems that have not been broken down into their component parts. For example, the automobile's fuel system, ignition system, and engine could be included in a single box, with stimulus provided by pressure on the accelerator, and response measured by the velocity of the vehicle. However, we could just as easily divide this box into several separate mechanisms (boxes) - engine, carburetor, distributor, etc. - each with its

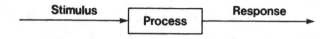

Fig. 1.4 A generalized block diagram

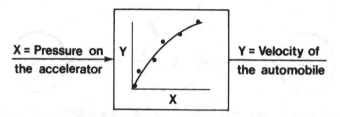

Fig. 1.5 An empirically defined process or "black box"

own stimulus and response. Thus, whenever we represent a complicated mechanism as a box we are confessing a lack of interest in or knowledge of the details of that mechanism and displaying more interest in the relationship between the stimulus and response variables. Because the details of the internal workings of the box are suppressed, they are frequently referred to as "black boxes" and are often described by rather simple empirical equations. For example, we can describe the process causing the automobile's velocity to change by measuring its velocity at several different accelerator depressions and then drawing a line through these sample points (Figure 1.5). This simple relationship substitutes for the complex real-life mechanisms of engine, carburetor, etc., and reduces the detail considerably. Reductions of this sort are often essential when we have to deal with extremely complex systems.

Mechanisms or processes may cause the value of the response variable to increase in direct relationship to inputs from the stimulus variable. This is called a positive process (+) and is illustrated by Figure 1.5; that is, increased pressure on the accelerator results in increased velocity and *vice versa*. In contrast, when the response variable changes in inverse relationship to the stimulus we have a negative process (−). For instance, *increased* pressure on the brake causes a *decrease* in velocity. Of course it is possible for a process to produce a constant response from a changing stimulus; the voltage regulator produces a constant voltage output from a variable voltage coming from the alternator. As we shall see later, such processes deserve our special attention.

The portrayal of a particular real-life system as a series of processes or mechanisms (boxes) linked together by variables (arrows) to produce a block diagram becomes our abstract *model* of the system we are investigating. The model is a simplification of the real system, with the fine details condensed into boxes and the larger enveloping systems relegated to the environment. The overall dynamic behavior of this abstract system is driven by inputs from the environment and its component processes, and this behavior is measured by changes in the state variables and the output variables (the arrows). The overall *qualitative* dynamics of the

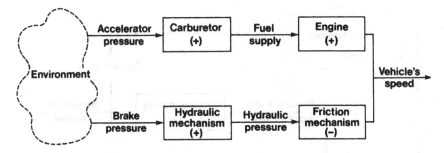

Fig. 1.6 Model of an automobile's power and braking systems

system can be determined by multiplying the signs of the component processes. For example, in the model of an automobile (Figure 1.6), pressure on the accelerator directly stimulates gasoline flow (+), which then directly affects the speed of the vehicle (+). The product of these two positive mechanisms is an overall positive effect of accelerator on velocity: (+)(+) = (+). In contrast, pressure on the brake pedal has an overall negative effect on velocity because we have the product of a positive and a negative mechanism: (+)(−) = (−).

1.5. Feedback Control

The systems we have considered so far are rather uninteresting because their dynamic behavior is completely determined by inputs from their environments. In our automobile example (Figure 1.6) the engine and braking systems are simply mechanisms for executing the orders of an environmental dictator (you). Systems become much more interesting and meaningful when they contain a degree of self-determination or internal control. A system may affect its own behavior when the output from a particular process feeds back to become the input for that same process at some time in the future, creating what is called a *feedback loop*. For example, if we include a driver in our automobile system then we will create a feedback loop composed of driver, engine, and speedometer (Figure 1.7). The driver (you) is now considered as a component of the system. You operate the vehicle by *comparing* your speed, provided by the speedometer, with the desired speed, obtained as an input from the environment (e.g., the posted speed limit). When the estimated speed is less than the desired speed, you increase pressure on the accelerator, and *vice versa*. Since you react in inverse proportion to the compared variables, you can be considered as a negative mechanism. Thus the feedback loop created by including the driver in the system has an overall negative effect because the product of the component processes is negative: (+)(+)(−) = (−). Negative feedback loops have very important effects on the dynamic behavior of a system because they tend to produce constant, or at least consistent, responses in the output variable(s). In other

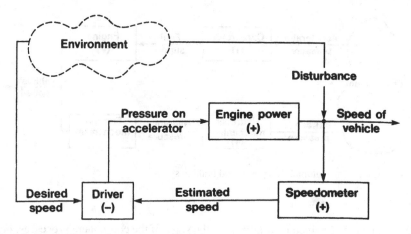

Fig. 1.7 Feedback control of an automobile

words, they tend to *control* the behavior of the output variable(s) and to iron out any disturbances to the desired system behavior. These disturbances are compensated for by internal adjustments of the various mechanisms. Let us use the automobile model to demonstrate the important attributes of systems with internal feedback control (Figure 1.7). Suppose that you are cruising at your desired speed when an outside disturbance, such as a downgrade, causes the speed of the car to increase. Looking at the speedometer, you realize that your estimated speed is greater than the speed limit and, acting as a negative process to oppose this divergence from your desired speed, you reduce pressure on the accelerator. The result of this negative feedback process is that the vehicle remains at, or close to, the desired speed at all times - its behavior is controlled. Negative feedback processes are found in most complex man-made and natural systems. The essential component is a mechanism that compares the actual behavior of the system with what is desired - the *comparator*. Familiar examples of comparators are governors, thermostats, autopilots, and the like. The analogue of the comparator is sometimes difficult to find in biological systems, particularly populations, communities, or ecosystems: What is the desired population density of a given species, say *Homo sapiens?* However, we will see later that negative feedback often occurs in populations and communities composed of living organisms, and that a comparator may not be involved in such systems.

The antipathy of negative feedback is positive feedback, which connotes lack of control, or the "vicious cycle" illustrated by the arms race in Figure 1.8. In this system all the processes are positive and the output, in terms of weapons deployment, tends to escalate with time. For example, if country A starts the "vicious cycle" by deploying a few offensive weapons, it is perceived as a threat by country B which then deploys weapons of its own, which is then perceived as a threat by A, and so on. Positive feedback, therefore, tends to amplify an initial movement or disturbance in the system's output. Although positive feedback was responsible for continual growth in weapons deployment in this example, it can also work in the

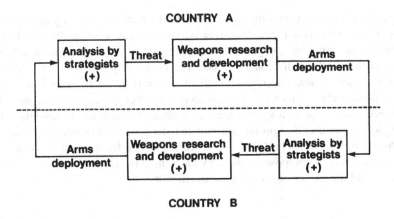

Fig. 1.8 The "arms race," a positive feedback system

opposite way and cause the system to decay continuously. For instance, if one country decreased its deployment of offensive weapons, then, according to the system we have depicted, the other country would be less threatened and would decrease its deployment, and so on until no more weapons existed. Once again, a movement in one direction is continuously amplified in the same direction as the initial movement. Because the positive feedback vicious cycle has to be initiated by someone, the initial disturbance is often cited as the cause of the problem, with cries of "they started it." However, the initial move cannot be amplified in the absence of a complete positive feedback loop and, if the loop exists, something will eventually set it off. It takes at least two to fight and two to make love and both are positive feedback processes. Thus, the structure of the loop is of more significance in the behavior of the system than the original move, which sets it off.

We can also see from Figure 1.8 how easily a positive loop can be changed into a negative one. For example, if one country decided to respond to a threat by reducing its arms deployment, this would change the sign in one of its boxes to negative and the total system to negative feedback, $(+)(+)(+)(-) = (-)$. An increased threat from its neighbor would now result in decreased weapons deployment, which would lower the threat to the other country. According to our diagram the other country would then reduce its arms deployment, lowering the threat to its neighbor. However, a lowered threat would cause this country to increase its weapons deployment and we can see that the system will remain at, or oscillate around, its original position.

In summary, then, positive feedback is a self-enhancing process in comparison to the self-controlling properties of negative feedback. In systems dominated by positive feedback we should expect very large effects building up from very small initial causes. Although some of these self-enhancing processes may be self-destructive, as implied by the terms *vicious cycle, arms race, inflation spiral, population explosion,* and the like, they need not necessarily be so. The agricultural revolution, knowledge explosion, and organic evolution are also positive feedback processes.

Feedback loops may pass through complicated pathways and many mechanisms before they return to their start. However, we can discover whether the total loop is positive or negative by applying the multiplication rule. Positive feedback will occur whenever all serially connected boxes in a loop are positive, or when an *even* number of them is negative (remember that $(-)(-) = (+)$). On the other hand, negative feedback only occurs when there is an *odd* number of negative processes in a loop. Feedback loops in a system may arise through design or circumstance. For example, the engineer designs the automobile to be controlled by negative feedback between driver and vehicle. On the other hand, positive feedback in the arms race was created, with no purpose implied, by the mutual interaction between two rival systems, and the circumstances of their interaction.

1.6. The Stability of Systems

A system is considered to be stable if its state variable(s) tend to return to or towards some particular *steady state* following an environmental disturbance. For this reason, stable systems are seen to persist over time in a state of balance, or equilibrium, with their environments. Thus television sets and automobiles perform consistently well because their designers were concerned with their properties of stability.

The concept of stability is extremely important to our understanding of dynamic systems and, perhaps, we can illustrate it with the example of a ball resting on different landscapes (Figure 1.9). In the first diagram (Figure 1.9A) the ball resting in the valley is in a stable state because it rolls back to the bottom of the valley following a disturbance. On the other hand, the ball on the mountaintop is in an unstable state because, if it is moved, it will continue to roll away from its original position (Figure 1.9B). The ball on the flat surface is said to be neutrally stable because it will remain wherever it is placed (Figure 1.9C).

At this point we need to distinguish between two kinds of stability. Systems are said to be *globally* stable if they return to their equilibrium position following a displacement of any magnitude, whereas those that only return if the displacement is relatively small are said to be *locally* stable in the neighborhood of the equilibrium point. For example, if the valley in Figure 1.9A was infinitely large, the ball would always return to equilibrium no matter how far it was moved up the walls of the valley. In this case the system would exhibit global stability. However, it might be more usual to find the landscape consisting of peaks and valleys, such as that shown in Figure 1.9D, in which case the system is only locally stable to a certain range of disturbances. When we make the landscape even more complicated we may find several locally stable equilibrium positions separated by unstable peaks (Figure 1.9E). These peaks, in actuality, define *thresholds* that separate the domains of different equilibria. For example, the ball in Figure 1.9E is sitting on the unstable threshold separating the domains of two equilibria, for a slight push one way or the other will result in its movement to one of these two positions. These concepts of local stability, multiple equilibria, and thresholds will prove to be very important later on in Chapters 3 and 4.

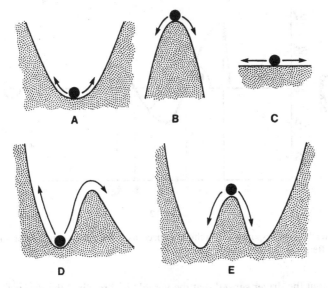

Fig. 1.9 Stable (A), unstable (B), and neutrally stable (C) landscapes, and locally stable landscapes with one (D) and two (E) equilibrium positions

As we might expect, the stability properties of a system are determined, to a large extent, by its feedback structure. When positive feedback loops dominate we will usually observe unstable growth or decay behavior and, sometimes, unstable thresholds. On the other hand, negative feedback loops will tend to control, or regulate, the system so that it performs in a consistent manner. They define the equilibrium structure of the system. Although stable systems are usually dominated by negative feedback control, we will see below that negative feedback is not a sufficient condition for stability.

The *dynamic stability* of systems governed by negative feedback can be evaluated by observing the behavior of the state variable(s) following a disturbance of the system from its steady state, under the condition that all environmental inputs remain constant. This is usually referred to as the system's *steady-state behavior*. Let us examine the steady-state behavior of the automobile-driver system illustrated in Figure 1.7. It will be in steady-state equilibrium when the vehicle is traveling at the desired reference speed, say 55 miles per hour. If an environmental disturbance causes a change in the vehicle's speed, the driver is notified by the speedometer and *compensates* for the disturbance by adjusting his pressure on the accelerator. A detailed examination of this process shows that it occurs in a series of steps through time (Figure 1.10). Suppose the driver notices an increase in his speed at time t_0 and responds by lifting his foot from the accelerator. The automobile will slow down and at time t_1 the driver will observe that the reference speed has been reached. He will then increase pressure on the accelerator in an attempt to maintain the desired speed. However, in the instant of time required to carry out these mental calculations, and for his reaction to be transmitted to the engine, the vehicle's speed will have dropped below 55 miles per hour. What we have here is a *time delay* between

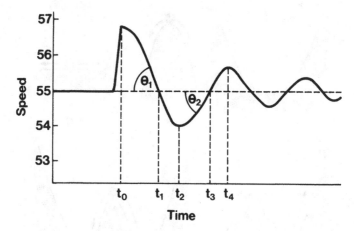

Fig. 1.10 Steady-state response of a vehicle's speed as it returns to equilibrium with damped oscillations after a displacement from equilibrium

the instant that the driver sensed that the vehicle had reached the desired speed and the time at which the engine responded with additional power. This time delay caused the actual speed to *undershoot* the desired speed. We might also expect that, after more power is given to the engine, the speed may *overshoot* the desired condition for the same reason. Thus, the speed of the car will tend to oscillate around its reference point, or equilibrium position. If the driver is able to improve his control with time, these oscillations will become smaller and smaller until the vehicle eventually attains the desired speed (Figure 1.10). If we examine this figure more carefully we will see that the size of the oscillations, given a constant time delay, depends on the angle of approach to the equilibrium line (i.e., θ in Fig. 1.10). This angle is a measure of the rate at which the car approaches the reference speed. We can see that, if this rate of approach decreases with time, then the oscillations will dampen out. This kind of steady-state behavior is usually called an approach to equilibrium with *damped oscillations*, and the system is said to be *damped stable*.

Two very important concepts have been introduced in the above paragraphs. The first is that delays in the negative feedback response may cause the system to overshoot its equilibrium position and exhibit oscillatory behavior. The second is that the degree of overshoot, and therefore the amplitude of the oscillations, is directly proportional to the length of the time delay and the rate at which the system approaches equilibrium.

The system that exhibits damped oscillations is, by definition, stable because it eventually returns to its steady state position. However, if the time delay is too long, or the rate of approach too fast, then the system can become unstable. For example, consider the case where the driver overreacts to a slight increase in speed by jamming on his brakes, causing the car to decelerate rapidly. His speed will undoubtedly undershoot the reference speed by a large margin (Figure 1.11). If the driver then flattens the accelerator in an attempt to regain his desired speed as quickly as possible, then an even larger overshoot may result. Continued *overcompensation* by the driver will cause

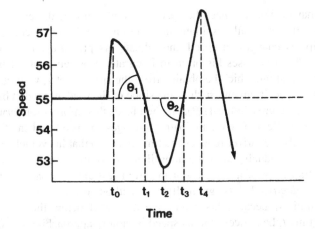

Fig. 1.11 The results of overcompensation for a disturbance in the speed of a vehicle

the oscillations to increase in amplitude and he will lose control of the car. The system, of course, is now unstable and the condition is usually referred to as *oscillatory instability* to distinguish it from the type of instability characteristic of positive feedback loops. Oscillatory instability results because the negative feedback processes overcompensate for the displacement from equilibrium caused by the initial disturbance. Figures 1.10 and 1.11 show that the degree of control that a driver has over his vehicle depends on the fineness with which he regulates acceleration and braking, as well as on his reaction time. Hence the advice of the driving instructor to use firm but gentle pressure on the pedals, and the admonishment against drinking while driving which dulls the brain and increases the time delay in the negative feedback response.

It is important to realize that, although negative feedback structures are designed to maintain a system in equilibrium, continuous environmental disturbances may prevent it from ever attaining the precise equilibrium point. No matter how finely you control your automobile it rarely remains for long at the precise speed you want because external conditions of wind, terrain, etc., change continuously. Hence, although equilibrium speeds certainly exist in the mind of the driver, they almost always deviate to some extent from this abstract reference point. Likewise, although we will rarely observe biological systems in precise equilibrium, we will frequently observe their tendency to return toward a particular state following environmental disturbances. Such tendencies should remind us that negative feedback processes are in operation.

1.7. Anticipatory Feedforward

We have seen that negative feedback loops can become unstable if the transfer of information, material, or energy through the loop takes a long time. The time delay can be reduced if the system contains a mechanism for anticipating, or predicting, its future behavior. For instance, the speed of an automobile driven by an experienced driver will not usually oscillate much around the desired speed because the driver anticipates changes in speed and adjusts his pressure on the accelerator accordingly. The driver uses his brain to integrate information about his present speed and acceleration, which he obtains from the speedometer, with observations from the external environment, such as the slope of the road, to predict his speed at some time in the future (Figure 1.12). He then feeds this information *forward* to his control of the accelerator. By anticipating changes in speed and making adjustments accordingly, the driver reduces the time delay so that his vehicle approaches the desired speed gradually and without oscillation. For instance, Figure 1.13 shows the velocity trajectory of a vehicle starting from rest and approaching its desired speed *asymptotically;* that is, gradually and without oscillation. The driver has accelerated initially because his actual speed is well below the desired speed. However, at time t_1 he notices that his speed is rapidly approaching the speed limit and he relaxes pressure on the accelerator in anticipation of reaching this speed. At time t_2 he predicts that he will not attain this speed unless he gives the car more gas, and reacts accordingly. At time t_3 he again anticipates reaching the correct speed and relaxes his foot, this time settling gradually into his desired equilibrium velocity. This negative feedback system, which now contains feedforward anticipation, is asymptotically stable because it approaches equilibrium without oscillation.

The critical component of a system with anticipatory feedforward is a predictive mechanism, or a model of how the system will behave under various environmental conditions. The experienced driver has a model in his mind of how the car will perform under different terrain and weather conditions, the model being constructed from past experiences. In a similar vein models of natural populations can be used by the manager to anticipate future population trends and to adjust his management plans. In a way the population manager is much like the driver of an automobile in

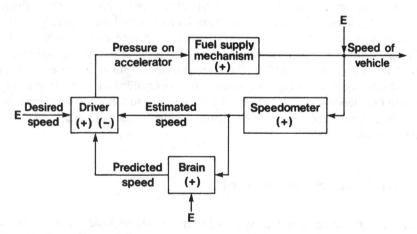

Fig. 1.12 Control of a vehicle's speed with negative feedback and anticipatory feedforward; the E's are environmental inputs

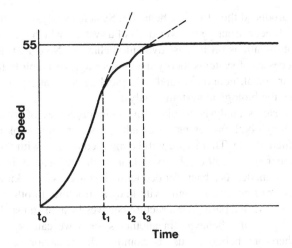

Fig. 1.13 Asymptotic approach of a vehicle to the desired reference speed

that he uses census estimates of the present population, with experience from the past built into a mental or mathematical model, to determine harvest levels. In this way he maintains a much finer degree of control over the population he is managing and minimizes any oscillatory or cyclic instability in the system.

1.8. Systems Analysis in Biology

The theory of dynamical systems was advanced, primarily by engineers, for designing complicated electronic and mechanical systems. In the mind of the engineer there is a picture of how the system should behave, a model if you like, and he designs the system to fulfill this concept. Thus, the best test of the engineer's competency is the actual performance of the system he designs. Control theory, particularly the concepts of negative feedback stability, serve as keystones in the design of dynamical mechanical and electronic systems. The success of dynamical systems engineering in such things as the space program attests to the power and utility of these basic concepts. Whether they are equally useful in biology is a question that the reader will have to decide for himself.

The investigation of complex natural systems is, essentially, a reversal of the engineering problem. Here the system already exists and the investigator is mainly interested in how it works. In other words, he is trying to understand why the system behaves as it does and to create a dynamical model of the workings, either in his mind or as a set of mathematical equations. His understanding comes by observing the behavior of the system as it responds to various environmental inputs, which may be natural or induced by the investigator. He then tries to deduce why the system behaves as it does; that is, he attempts to deduce the characteristic structure, or design, which produced the observed behavior. Systems analysis in biology, therefore, is the *art* of reconstructing the workings of a system, which the analyst did not design, from observations on its past dynamic behavior (Note 1.2). In contrast to the striking successes of systems theory in engineering, our inability to understand and manage our social, economic, and biological systems attests to the difficult problems facing the biological systems analyst.

Biological systems analysis involves the twin processes of observation and deduction. Although both processes are equally important, this book leans heavily toward the deductive side. That is, we will be more concerned with the structure of systems that other investigators observed than with the manner in which those observations were made. As a basis for deduction it is necessary to know something about the behavior of general systems with known structure. In other words, if we know that a system with a particular structure behaves in such and such a way, then when we observe similar behavior in another system we can propose a similar structure. It therefore behooves us to examine the behavior of some simple systems.

The first, and perhaps most important observation, is that systems with rather simple feedback structure and obeying simple rules often exhibit an astounding array of dynamic behavior. This property can be demonstrated using the so-called "Game of Life," invented by the mathematician John Horton Conway (Note 1.3). The game is played on a large checkerboard, and the pieces (checkers) represent living organisms. The birth of new individuals and the death of old ones is governed by three simple rules: Every "organism" with one or less neighbors dies from isolation; every one with four or more neighbors dies from overcrowding; and a new individual is born to any empty square that is adjacent to exactly three "organisms." We can see from Figure 1.14 that there are three feedback loops: two positive and one negative. This might lead us to expect growth, decay, and equilibrium as possible behavioral patterns in the dynamic repertoire of the population.

The game is played by positioning a few counters on the board in a particular pattern, and then observing how the size and pattern of the population changes through time as the rules are applied. Figure 1.15 shows some numerical patterns that were produced when we started with six "organisms" arranged in three different starting configurations. As you can see, the dynamics were considerably different: Population A attained a steady state of four individuals after only one move,

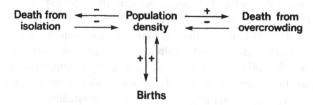

Fig. 1.14 Feedback structure of the "Game of Life"; the signs of the processes (arrows) show the qualitative effect of one state variable on another, so that increased population density causes increased deaths from overcrowding (+) but decreased deaths from isolation (−)

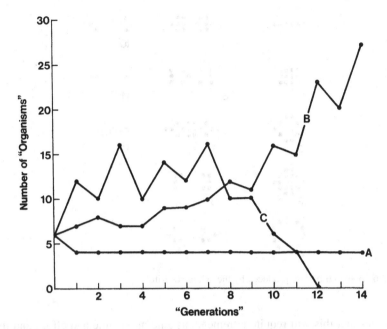

Fig. 1.15   Three numerical patterns produced by the "Game of Life"

population B increased slowly for nine moves and then grew quite rapidly in a series of jumps, while population C oscillated for eight moves before declining to extinction. We could continue to produce a large number of similar simulations, but we would come to the same conclusion; namely that this system produces a confounding array of dynamic patterns in space and time, and that these differences are purely a product of the starting pattern and not of any internal changes in the structure of the system or its environment.

The game also illustrates how we can improve our understanding of a system by examining the behavior of additional state variables. For example, it is difficult to explain why population C in Figure 1.15 became extinct by examining its numerical dynamics alone. However, if we look at its spatial pattern the cause of its demise becomes apparent (Figure 1.16). Here we see that the center of the population became very overcrowded in generation 7. This overcrowding caused high mortality in the center, which resulted in two separate subpopulations in generation 8. These populations were too sparsely distributed to maintain growth and they died out from the effects of isolation by generation 12.

Although the "Game of Life" is but a parody, and should not be confused with real-life systems, it does show us that feedback systems governed by very simple rules can exhibit a confounding array of dynamic behavior. We can see that it may be difficult to understand the internal structure and processes of a system from empirical observations of its dynamics alone. It may be possible to describe most of the patterns that a system exhibits by observing it under a large number of different

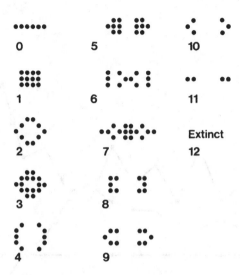

Fig. 1.16 A spatial pattern produced by the "Game of Life"

conditions, but this will require a tremendous amount of time and effort, and there will be no assurance that all possible behaviors have been observed. Thus, the weakness of the empirical approach is that predictions cannot be made with any confidence unless the system has previously been observed operating under similar conditions.

An alternative approach is to try to understand and describe the structure and processes of the system. The amount of information required to do this is usually much less than is needed to describe its complete array of dynamic behavior and, *if* the system is defined accurately, it will accurately predict behaviors that were not previously observed. However, this approach requires a considerable amount of intelligent detective work and judgment on the part of the analyst when trying to unravel the intricate network of interactions and interdependencies that make up the internal structure of a complex system.

The detective and the biosystems analyst have much in common. The detective attempts to reconstruct, from a series of clues, the probable chain of events that led to a particular crime. His deductions are made possible because he has a general understanding of human nature and, in particular, the criminal mind. The systems analyst works in a similar fashion. His clues are the behavior he observes in certain state variables as the system changes in response to environmental conditions. Based on these observations, and with a general understanding of how systems with known structure behave, he deduces the probable structure of the observed system. He then builds a model of the system "as he sees it" and evaluates it by comparing its predictions under given conditions with that of the real system operating under the same (or similar) conditions.

Our general understanding of feedback loops and how they affect the dynamic behavior of systems is particularly useful. We know that negative feedback loops frequently induce steady-state behavior or oscillatory instability. On the other hand, positive feedback loops usually cause exponential growth or decay dynamics. For example, in the "Game of Life" dynamics illustrated in Figure 1.15 population A exhibited steady-state behavior, B a growth process, while C decayed to extinction. From these few observations we might deduce that the system contained *at least* three feedback loops: a stabilizing negative loop, a positive growth loop, and a positive decay loop. As we know, these correspond to the "death from overcrowding," "birth," and "death from isolation" processes shown in Figure 1.14. Of course, systems of intercommunicating feedback loops may have much more complicated behavioral patterns than those discussed above. For instance, population C in Figure 1.15 oscillated for eight generations before it started on its path to extinction. This may give us a clue that time delays are present in the negative feedback structure. A time delay is present in the "Game of Life" because the numbers at one point in time (after a move) depend on the numbers and spatial distribution of organisms at the beginning of the move.

The deductive process leaves the systems analyst with a concept in his mind about the design or structure of the system he is observing. In order to discover whether this conceptual model is the correct one, he must formulate it as a quantitative model and test its predictions against new observations. The model structure may be represented as a block diagram or flow graph composed of state variables and their mathematical linkages (processes). The dynamic behavior of this model can then be compared with real-life observations made under similar operating conditions. If the model fails to behave like the real system, the deductive arguments are assumed incomplete or inaccurate, and the analyst has to refine his concept of the system. The process of evaluating the behavior of a model by comparing it with the range of behavior observed in the real world is known as *validation*, or, more correctly, *invalidation*. An invalid model means that the analyst must return to square one and again observe the behavior of the natural system. He may have to collect new data or evidence and then deduce a new structure and equations to explain all the observations he has made. Through the repetitive process of observation, deduction, and invalidation, the model is slowly refined until it simulates the behavior of the real system in a manner that satisfies the analyst (Figure 1.17). Model building may be thought of as a feedback process in which the model is continuously improved to meet some predetermined qualitative or quantitative criterion; for example, the analyst may be satisfied if the model simulates the general qualitative behavior of the system (i.e., steady states, growth, oscillations, etc.), or may demand quantitative predictions with particular precision (e.g., the observed values do not deviate more that 10% from the prediction). At the end of this process the analyst should have a working concept of the system's structure, which should enable prediction of future behavior over a wide array of environmental and initial conditions. But the analyst remains in an unenviable position, for the model can never be proven correct. It can, however, be invalidated when observation are made that conflict with the prediction of the model. This situation should be kept in mind as we construct models of population systems later in this book.

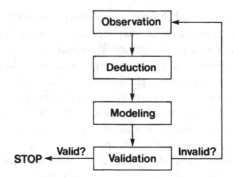

Fig. 1.17 The process of constructing a model of a system

1.9. Chapter Summary

The main points discussed in Chapter 1 are emphasized below:

1. A system is an assemblage of physical objects, parts, or components that interact or communicate with, or are interdependent on each other so as to operate as an integrated whole. The extent or boundary of a particular system is defined by the interests and perspective of the observer.
2. All systems exist in space and time within a larger universe as part of a hierarchy of systems. This enveloping universe is called the system's environment, and it supplies all the material, energy, or information necessary to make the system run, or which may disturb or destroy it.
3. A system can contribute material, energy, or information to its environment, which may cause the environment to change to the benefit or detriment of the system.
4. The state of a system at a particular time and place is described by the condition of its state variables.
5. The dynamic behavior of a system describes the changes that occur in its state variables in time and space. Changes in state may be caused by a state variable's interaction with its environment, with other state variables, or with itself. The resultant of all interactions may cause the state variable to increase (+), decrease (−), or remain unchanged (0).
6. Block diagrams are composed of boxes, which represent processes, and arrows, which represent variables. Processes are stimulated by inputs or state variables and produce responses in outputs or state variables.
7. Positive processes or mechanisms produce responses that are directly related to the stimulus, while negative processes produce responses that are inversely related to the stimulus.
8. In chains of linked processes the overall stimulus-response relationship can be determined by multiplying the signs of the component processes.

9. When a response is transmitted back to determining process, even if it passes through a number of intervening processes, a feedback loop is created.

10. Negative feedback exists when the product of the signs of all processes in a feedback loop is negative, and positive feedback exists when the product is positive.

11. Positive feedback loops usually amplify an initial stimulus or disturbance. The state variables move in the same direction as the initial stimulus so that they either grow or decay continuously.

12. Negative feedback loops usually attenuate or dampen an initial stimulus or disturbance so that the state variables tend to return towards their original conditions. In contrast to positive feedback, negative feedback loops often stabilize the dynamics of a system.

13. The degree of stability induced by a negative feedback loop depends on the speed at which the response is transmitted back to its source, and the vigor of the negative, compensatory processes in the loop. That is, fast-acting gentle mechanisms induce greater stability than slow-acting harsh processes.

14. When information concerning the expected behavior of a state variable is fed forward to the control mechanism in a negative feedback loop, a greater degree of control and stability is possible. Feedforward anticipation involves the prediction of future system behavior from its present state and observation of environmental conditions.

15. Natural systems can be analyzed by (a) observing the behavior of the system under an array of environmental conditions; (b) deducing the structure (boxes and arrows) of the system, particularly the feedback loops; (c) constructing a model of the system from the deductions; (d) evaluating whether the model behaves in a manner similar to the real system; and (e) returning to (a) if the model is unsatisfactory.

Exercises

1.1. In winter, the temperature of a room is controlled by a thermostat linked to a furnace. Draw the structure of this system using a block diagram and describe the feedback loop. An experiment was performed to measure the actual temperature in the room with a thermometer and it was found that the temperature cycled around the thermostat setting. Explain the probable cause of these cycles.

1.2. Insert the disk that comes with this book in your computer and follow the instructions to find the program that simulates the "Game of Life". Start with 5 individuals and run simulations in different starting configurations. Explain the dynamic behavior you observe and the causes of that behavior. Repeat with different numbers of starters. What is the single most important conclusion from this exercise? For those interested in more information on the game, search the internet under "game of life".

Notes

1.1. For a general discussion of dynamical systems theory and its application in the biological sciences, the student is referred to the following works:

- *Positive Feedback—A General Systems Approach to Positive/Negative Feedback and Mutual Causality*, edited by J. H. Milsum, published by Pergamon Press, New York, 1968, is a compilation of works that examines the philosophical, historical, and technical aspects of dynamical systems theory, and its application in the biological and social sciences, in a manner comprehensible to the general scientific community.
- Biological Control Systems Analysis, by J. H. Milsum, published by McGraw-Hill Book Company, New York, 1966, is a much more technical treatment of dynamic systems theory for the advanced student. Although it is largely concerned with physiological systems, some population concepts and their control theory analogues are introduced. The general reader may find Chapters 1 and 2 a useful, if rather technical, introduction to dynamic systems and their control.
- *Feedback Mechanisms in Animal Behavior*, by D. J. McFarland, published by Academic Press, New York, 1971, is, as the title indicates, mostly concerned with the application of control theory to behavioral systems. However, a lucid introduction to the elements of control theory is presented in Chapters 1 and 2.

1.2. We often tend to draw rigid distinctions between the arts and sciences when such distinctions are fuzzy, at best. Many scientists spend much of their time in what can only be described as artistic endeavors. This is particularly true of those involved in the analysis and synthesis of natural systems. The artist uses concrete materials to construct an abstract model of something that exists in his mind. Likewise, the systems analyst uses concrete scientific information to construct an abstract model of how he thinks the system works. The model is *his* conception and, therefore, its resemblance to reality is only as good as his facts and his innate abilities to synthesize those facts into a model of the system. The scientific method called the "hypothetico-deductive" (H-D) approach involves the validation, or better invalidation, of the conceptual model (see Stephen Fretwell's book *Populations in a Seasonal Environment* for a nice summary of the H-D philosophy applied to ecological problems; the book was published in 1972 by Princeton University Press as part of their series entitled *Monographs in Population Biology*).

Because the "art" of constructing abstract models rests on knowledge and insight concerning the nature of the system being analyzed, it is important that biological models arise in the minds of experienced and intelligent biologists. In the past, however, many biologists, although able to see the picture, were unable to paint it because they were unfamiliar with the tools - mathematics. Consequently incomplete or inaccurate pictures were often painted by those who were - the mathematicians. Fortunately this scene is slowly changing as biologists learn how to use the mathematical tools and mathematicians become students of biology.

1.3. The "Game of Life" was first reported in the Mathematical Games section of *Scientific American*, vol. 223, no. 4, October 1970. Since then it has become a popular game amongst schoolchildren as well as professors of mathematics. The game can be accessed through most computer systems, usually under the code name LIFE. We heartily recommend that you invest a few hours playing the game to obtain a feel for the rich variety of dynamic behavior that can result from the application of even the simplest rules in a feedback structure. However, beware! People have become addicted to this game and withdrawal may be painful.

Chapter 2
Population Dynamics and an Elementary Model

2.1. What is a Population?

We can think of a population as a group of individuals of the same species, which live together at the same time and in the same place. This statement implies the coexistence of, and potential interaction or intercommunication between, all the members of the population, and that the population is distinctly defined in space. Space should be tied in to the biology and behavior of a species – for example, an acre is too small to study an elk population. The spatial element, which is implied in statements such as "the population of New York" or "the population of insects in a wheat field," is very important because it delimits the geographic boundaries of the population system being considered. Although the boundaries are often drawn rather arbitrarily, they should, ideally, enclose a distinct population unit (a much more strict definition used by systematic biologists is presented in Note 2.1).

The members of a population may interact in a number of ways. They may cooperate with each other during certain activities, such as hunting or nest building. At other times they may compete with each other for essential resources, such as food or space, which are in short supply. Of course, individuals also mate with each other to reproduce new individuals. As a result of these interactions new individuals are born into the population whilst others are lost.

The environment surrounding the population provides it with resources, such as food and shelter, as well as pressures from predators, parasites, and competition with other species of organisms. Immigrants may also enter from other nearby populations or individuals may emigrate out of the population.

These ideas concerning the structure and functioning of a population system are summarized in Figure 2.1. Although this scheme may be the most logical way to view the population as a dynamic system, it poses some severe analytical problems. In particular, each individual is treated as a separate component of the population, forcing one to consider the possible interactions between each individual and all other members. When a population is large, as many are, the number of potential interactions becomes astronomical, equal to $n(n - 1)$, where n is the number of individuals. Thus, a population of one thousand members will have almost a million potential interactions ($1000 \times 999 = 999{,}000$). In order to reduce

A.A. Berryman, P. Kindlmann, *Population Systems: A General Introduction*
© Springer Science+Business Media B.V. 2008

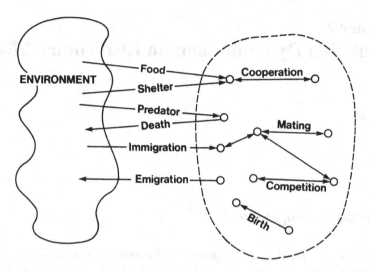

Fig. 2.1 The population as a group of interacting individuals of the same species coexisting within specific geographic boundaries in an interval of time during which certain discrete events, such as births, deaths, and migrations, occur

the number of calculations, we often assume that all members have equal opportunity to interact with each other and, in so doing, produce births (natality), deaths (mortality), and migrations, which are characteristic of that particular population. These characteristic processes will be determined by the average properties of the membership, and of the environment in which they are living, and their operation will produce changes in the state of certain population variables. These ideas are summarized in Figure 2.2, where the average individual properties, acting with the environment, control the processes of population change which, in turn, affect certain population state variables, such as density, spatial arrangement, age distribution, or the frequency of certain genes. Feedback loops may be formed if the state variables affect the properties of individuals or if they influence the environment. For example, dense populations may cause increased movements amongst certain individuals, resulting in emigrations, which may lead to changes in the structure of the population; that is, certain age groups or genotypes may emigrate whilst others remain. Dense populations may also affect their environments when waste products accumulate (pollution) or resources such as food and nesting sites are exhausted.

Our view of the population as a number of individuals with an average set of properties may leave some, including the authors of this book, with an uneasy feeling. The qualities of individual choice and action have been suppressed for the purpose of simplicity and tractability. However, until a systematic approach is developed which permits the expression of individual action, without the necessity of considering all possible individual variations and interactions, we must be satisfied with our present concept, or throw up our hands in despair.

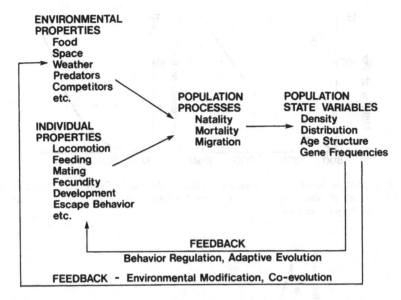

Fig. 2.2 The functioning of a population system

2.2. Dynamics of Populations

In Chapter 1 the investigation of population systems was likened to a problem of detection. By observing the behavior of the system, searching for clues, and then using our basic knowledge of general systems dynamics, we can often deduce the probable structure that produced the behavior we observed. In this vein, let us now look at some characteristic patterns of behavior, which have been observed in natural populations. The analyst will first observe that populations can exhibit a confounding array of behavior. Remembering our experience with the "Game of Life," we know that even simple rules may produce complex behavior when feedback loops are present in the system. Therefore, let us look at population dynamics with an eye for the possible feedback structure that produced the observed behavior.

In these times of dwindling natural resources we are all aware of the phenomenon of population growth. Some populations, such as that of our own species, give the impression of continual growth (Figure 2.3A). Patterns of this kind are more commonly observed when species are colonizing a new and favorable environment. A typical example is the growth of a pheasant population after its introduction into an island off the coast of Washington (Figure 2.3B). The same population growth pattern is typical for insect pests during their outbreaks. From these observations we might deduce that population systems contain a positive feedback loop that enables them to grow when environmental conditions are favorable.

In contrast, some organisms have declined to eventual extinction; examples are the dodo, passenger pigeon, and dinosaur. The blue whale population illustrated in

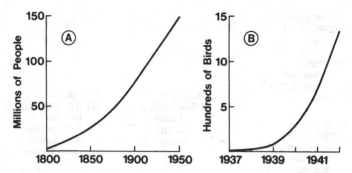

Fig. 2.3 Population growth of (A) the human population of the United States (U.S. Bureau of the Census), and (B) pheasants on Protection Island, Washington (redrawn from A. S. Einarsen. *Murrelet*, vol. 26, pp. 2 and 39, 1945)

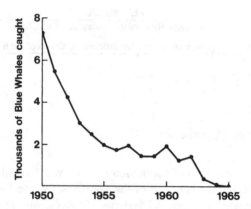

Fig. 2.4 Catch records of blue whales from the *Yearbook of Fishing Statistics*, Food and Agriculture Organization of the United Nations, Rome

Figure 2.4 may be heading for a similar fate. Although the extinction of species may be a matter of grave concern, it is a relatively rare event in the time scales with which we will be concerned. However, it is fairly common to observe the decline and extinction of populations in particular localities. These local extinctions are usually observed when the environment in a particular area becomes very unfavorable for the species, either through severe natural alterations of the physical conditions, the destruction or pollution of the environment by the population itself, or the actions of other organisms such as man. The pattern of decline illustrated by Figure 2.4 indicates the presence of a positive feedback loop because the population continues to change in the direction of the initial movement. As both growth and decline patterns seem to be associated with properties of the environment, we might suspect that they are controlled by the same feedback loop, and that the environment determines, which pattern is exhibited. Thus, we should expect populations to grow in favorable environments and to decline in unfavorable ones.

Although populations may exhibit growth and decline patterns over certain periods of time, they eventually reach a condition of equilibrium with their environments. This equilibrium is usually attained at some positive population density, as is demonstrated by the barnacle population in Figure 2.5. However, equilibrium at zero density, or extinction, is always a possibility. Many populations that we observe seem to remain for long periods at relatively constant densities, or to oscillate around some characteristic density. For example, the population of hole-nesting songbirds shown in Figure 2.6 fluctuated consistently around an average density in both oak and pine woods, although the mean population level was much higher in the former. It seems, therefore, that populations must be influenced by a negative feedback loop, which tends to regulate them at some characteristic density. Moreover, the level of regulation, or characteristic density (in systems terminology the "desired" or "reference" level), appears to be determined by environmental properties. If this is true, then the characteristic density should change if the environment is altered. The experiment illustrated in Figure 2.7 provides us with some confidence in this line of reasoning because a change in the environment brought about by thinning the forest resulted in a change in the great tit's characteristic density.

Let us now turn our attention to the oscillations seen in Figures 2.6 and 2.7. Although the bird populations seem to be regulated at a characteristic density, they fluctuate to varying degrees around this level. From our understanding of general systems dynamics we might suspect that these oscillations are due to minor environmental disturbances or to negative feedback mechanisms that act with a time delay. In fact, both factors are probably involved because small displacements from equilibrium caused by minor environmental fluctuations are necessary to initiate the negative feedback response.

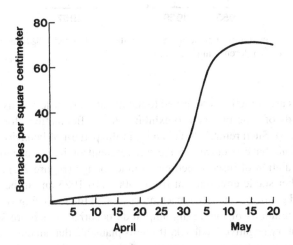

Fig. 2.5 Barnacle larvae settling on exposed rocks in the Firth of Clyde (redrawn from J. H. Connell, *Ecological Monographs*, vol. 31, p. 61, 1961)

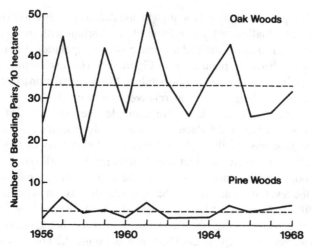

Fig. 2.6 Populations of breeding great tits in oak and pine forests in Holland (redrawn from H. N. Kluyver, *Dynamics of Numbers in Populations*, p. 507; Centre for Agricultural Publishing and Documentation, Wageningen, Netherlands, 1971)

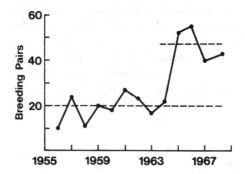

Fig. 2.7 Breeding populations of great tits in a wood that was thinned in 1963 and later (redrawn after H. N. Kluyver, see reference in Figure 2.6)

Populations are sometimes observed to remain at more or less constant densities for long periods of time but then to exhibit extreme fluctuations for a short time span (Figure 2.8). Such remarkable changes in the pattern of behavior suggest that severe alterations have occurred in the environment or in the negative feedback loop. The population of insects feeding on pine foliage (Figure 2.8) seemed to be living in a rather stable environment from 1900 until 1925 and to be regulated by fast-acting and gentle negative feedback processes. In the following years, however, a series of oscillations of increasing amplitude occurred which are suggestive of overcompensatory negative feedback. It seems plausible that an environmental disturbance caused a change in the properties of the negative feedback structure, precipitating the dramatic population behavior. For instance, the disturbance could

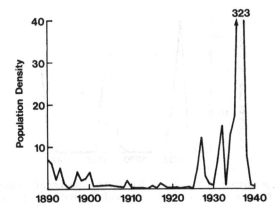

Fig. 2.8 Population fluctuations of a moth feeding on pine needles in Germany (redrawn from F. Schwerdtfeger, *Zeitschrift für angewandte Entomologie*, vol. 28, p. 254, 1941)

have disengaged the gentle, fast-acting mechanisms, which regulated the population at its previous low level. The released population may then have come under the influence of delayed feedback processes, which over-compensated for the density changes and produced the population oscillations. We will examine the possible causes for such divergent population behavior later in this chapter and in more detail in Chapter 3.

While irregular population explosions, which are often called outbreaks or epidemics, are characteristic of certain species, others seem to go through regular cycles of growth and collapse. There appear to be two general classes of population cycles: short cycles of a period of 3 to 5 years exhibited by lemmings and other small rodents inhabiting the artic tundra; and long cycles of 7- to 10-year periodicity that are characteristic of many forest insects, game birds, and larger tundra mammals (see Note 2.2). It is also interesting that the cycles of a particular species are often synchronized, or in phase with each other, even though the populations may be widely separated in space from each other. In addition, the cycles of different species also seem to be in phase in certain cases, such as several species of game birds in North America.

One of the most studied cyclic populations is that of the larch budmoth in the Engadin Valley of Switzerland (Figure 2.9). This insect defoliator of larch and pine goes through regular 9-year cycles high in the Swiss Alps. However, at low elevations the populations remain at relatively constant densities, fluctuating between 50 and 100 individual larvae per unit of larch foliage; this is in comparison to the 20,000-fold or greater increases and decreases seen in the high Alps (see Note 2.3). Similar phenomena have also been observed with other forest insects in Europe and North America, as well as with small rodent populations in Scandinavia and Germany. We are left with the general impression that populations can cycle in certain environments, but remain rather stable in others. Once again, the environment seems to play a decisive role in this difference in behavior.

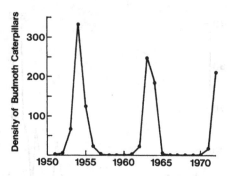

Fig. 2.9 Nine-year population cycles of the larch budmoth in the Engadin Valley of Switzerland (from the works of, and personal communication with, W. Baltensweiler; see Note 2.3 for reference)

From a systems viewpoint we know that cycles may be caused if time delays are present in the negative feedback loops, and if the feedback mechanisms exert strong control on the system. This observation may lead us to deduce a similar cause for cyclic population dynamics. It is apparent that the environment plays a decisive role in maintaining or suppressing these cycles and may also be important in synchronizing them. The latter conclusion is based on the observation that population cycles are frequently in phase over broad geographic regions, and sometimes even between different species of organism. This means that the forces involved in synchronizing the cycles must operate over extensive areas and affect different species similarly, which suggests that climate or weather are probably involved. For example, severe winter temperatures and deep snow may cause a drastic reduction in the populations of a number of species over a wide area so that they all start at the same low densities and thereafter cycle in unison. Therefore, catastrophic environmental forces, acting simultaneously on different populations, may operate to synchronize the cycles even though they only occur at rather rare intervals (e.g., once every 20 or 30 years).

We have now looked at a number of ways in which natural animal populations behave, in particular, population growth and decline, equilibrium behavior, irregular outbreaks, and population cycles. From this we have deduced that at least one positive and one negative feedback loop must be involved in most population systems, and that the environment plays a crucial part in determining whether growth or decline occurs, setting equilibrium levels, and influencing the stability of negative feedback loops. Of course, we have not covered the complete waterfront. Populations can be found that do not seem to fit into these general patterns. In particular, rather haphazard and violent behavior may be observed in populations inhabiting extremely variable environments. In addition we have restricted ourselves to observations on a single state variable – population numbers or density. This is because numbers are usually measured by population ecologists whereas spatial arrangement, age distributions, or gene frequencies are measured less often. We will have to consider these variables later in this book. Let us then proceed to the next step in systems analysis: the construction of a model.

2.3. An Elementary Population Model

A population was previously defined as a group of coexisting organisms of the same species, which, on interacting with each other and with their environment, give rise to changes in the abundance of individuals in the population. During any given interval of time, each organism may reproduce, die, or migrate into or out of the geographic region bounding the population. The sum of all these individual activities produces a change in the population, which can be expressed as

Population change = Births + Immigration − Deaths − Emigrations

or

$$\Delta N = B + I - D - E, \tag{2.1}$$

where ΔN represents a change in population density over a particular interval of time - usually a year or a generation. Now suppose that at the end of this period of time there were N_t individuals in the population, and that at the beginning, or at the end of the previous time period, there were N_{t-1} individuals. Then we can also express population change as

$$\Delta N = N_t - N_{t-1}. \tag{2.2}$$

However, as birth, death, and migration are things that happen to individuals, rather than populations, it is probably more meaningful to express population change in terms of the individual. We can obtain a measure of the rate of change in the average individual by dividing everything by the size of the population at the beginning of the time interval, so that

$$R = \frac{\Delta N}{N_{t-1}} = \frac{N_t - N_{t-1}}{N_{t-1}} = \frac{B + I = D - E}{N_{t-1}}, \tag{2.3}$$

where R represents the *individual, or per capita, rate of increase* over the time period $t - 1$ to t. We can see that when R is zero then births and immigrants exactly balance deaths and emigrants, and the population will remain at a constant level over the time period. However, when births and immigrants outnumber deaths and emigrants, R will be greater than zero and the population will grow, while *vice versa* the population will decline.

We can now write our simple model in terms of population size per unit area, N, and the individual rate of increase per unit of time, R, as follows:

$$R = (N_t - N_{t-1})/N_{t-1},$$

or

$$N_t = N_{t-1} + RN_{t-1} \tag{2.4}$$

From this equation we can calculate the density of the population at the end of a particular time increment from its density at the beginning of the interval and the individual rate of increase. The equation describes a positive feedback loop because the numbers at time t can be fed back into the right side of the equation to calculate population density in the next time period, provided of course that R remains constant; that is, N_t becomes N_{t-1} as we move into successive time periods. The structure of this positive feedback loop is shown in Figure 2.10.

To demonstrate the dynamic properties of this system, let us suppose that we have a population of 10 organisms at time zero and that the per capita rate of increase remains constant at 0.5 individuals per unit of time. At the end of the first time period the population will be

$$N_1 = N_0 + RN_0$$
$$= 10 + 0.5 \times 10 = 15$$

Using the new population as the input for the next cycle we get

$$N_2 = 15 + 0.5 \times 15 = 22.5$$

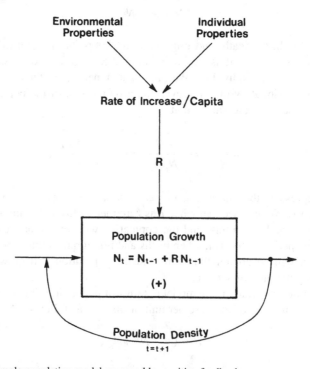

Fig. 2.10 A simple population model governed by positive feedback

If we continue through the feedback loop for two more time periods we will obtain the growth curve shown in Figure 2.11A. What we observe is the familiar exponential, or geometric, growth pattern, which is so characteristic of populations growing in a favorable and unrestricting environment (e.g., Figure 2.3). As we noted earlier, the system will only exhibit equilibrium behavior if R is zero (Figure 2.11B), and if R is negative, it will decline geometrically to extinction (Figure 2.11C), as may be happening to the blue whale population of Figure 2.4. In most ecological texts this population growth model is formulated as a differential equation, rather than the difference equation given by (2.4) (see Note 2.4). Note that when logarithms of population numbers are plotted on the vertical axis, the dependence in Figure 2.11 becomes linear, which is easier to analyze than exponential dependence. Indeed, instead of raw population numbers we often use their logarithms, which makes the analysis easier. We will see another example when we will be speaking about the logistic growth.

We now seem to have a fairly reasonable model of the positive feedback loop, which we deduced must exist in our population system. This loop will usually cause a population to grow indefinitely or to become extinct, both of which are rather unusual occurrences. Therefore, the positive feedback tendencies must be counteracted by one or more negative feedback loops. Although this was realized by a number of early philosophers, it was the English clergyman Thomas Malthus who, in 1798, produced the first definitive treatise on the subject. His book, *An Essay on the Principle of Population*, presented the view that, when populations become very

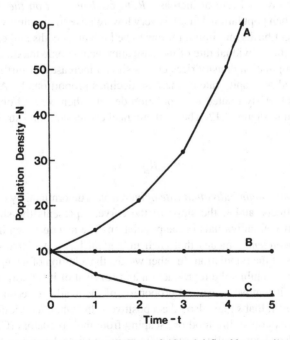

Fig. 2.11 Dynamics of the model shown in Figure 2.10 and defined by equation (2.4) when $N_0 =$ 10 and $R = 0.5$ (A), $R = 0$ (B), and $R = -0.5$ (C)

dense, there is an intense struggle between individuals for a diminishing supply of natural resources. In other words, the *demand* for resources by the expanding population must at some point exceed the *supply* and, when this happens, the members of the population will compete with each other for the diminishing supply of resources. Supply, demand, and competition are, of course, a fundamental principle in all animal and plant economies, not to mention the complex economic systems of human societies. It is not surprising, therefore, that these ideas arose in the mind of an economic philosopher such as Malthus.

Malthus argued that a balance between supply and demand could only be attained through changes in a population's variables; that is, changes in the birth, death, or migration rates. At the time these ideas bordered on heresy and gave rise to waves of theological and biological controversy, which have not completely subsided to this day. However, his ideas also stimulated a new school of population biologists. In particular, his concept of a "struggle for existence" led directly to Darwin's theory for the evolution of species, and played an important part in the thinking of the early mathematicians Verhulst, Lotka, and Volterra who played such an important role in population theory. For these reasons, that erstwhile clergyman, Thomas Robert Malthus, is considered by many as the father of population biology (Note 2.5).

If we accept the proposition that competition for a scarce resource is reflected by changes in the crucial population parameters of births, deaths, and migrations, then we can introduce the Malthusian arguments into our population model by allowing the individual rate of increase, R, to be *dependent on the density of the population*. When population density is very low, relative to the supply of resources, we would expect births and immigrations to be high and deaths and emigrations to be low so that the individual rate of increase approaches some maximum called R_m. However, as population density rises deaths should increase and births decrease so that the realized per capita rate of increase declines proportionally. Assuming that this decline is linearly related to population density, then we will obtain the relationship shown in Figure 2.12. The mathematical expression for a straight line with negative slope is

$$R = R_m - sN, \tag{2.5}$$

where R_m, the *maximum individual rate of increase*, is determined by environmental and genetic effects, and s, the slope of the curve, represents the strength of the interaction between individuals in the population. The negative sign in front of the *interaction coefficient s* means that each individual has a negative effect on the other members of the population. In other words, the addition of a single individual to the population inhibits the reproduction and survival of its cohorts by the quantity s, because it removes a certain proportion of the available resources. We will see later, however, that s may also take a positive value when individuals cooperate with each other in obtaining food or escaping from their predators (Chapter 3). Of course it is also theoretically possible for individuals to have no effect on each other, in which case s will be zero and, from equation (2.5), the realized individual

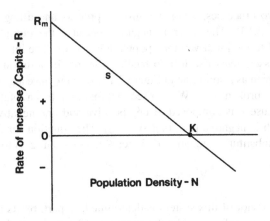

Fig. 2.12 A linear relationship between the individual rate of increase, R, and population density, N, where R_m is the maximum per capita rate of increase, s is the slope of the relationship, and K is the equilibrium density

rate of increase will equal the maximum $(R = R_m)$. Perhaps we can visualize this more clearly in the following statement:

$$
\begin{bmatrix} \text{Realized per} \\ \text{capita rate} \\ \text{of increase} \end{bmatrix} = \begin{Bmatrix} \text{Maximum per} \\ \text{capita rate} \\ \text{of increase} \end{Bmatrix} + \begin{Bmatrix} \text{Intensity of} \\ \text{interaction} \\ \text{between individuals} \end{Bmatrix} \times \{\text{Population density}\}
$$

$$
R \;=\; R_m + \begin{pmatrix} + \\ 0 \\ - \end{pmatrix} s \;\times\; N
$$

Of course when s is negative then this statement is identical to equation (2.5).

Because the interaction coefficient measures the intensity, or strength, of the interaction between individuals, it will assume a larger negative value in those species, which utilize large amounts of the limiting resources. For this reason we would expect that, given equal resources, large organisms (elephants and whales) will have larger interaction coefficients than smaller ones (insects and crustaceans).

The Malthusian concept of a struggle for existence is expressed in the quantity sN of equation (2.5). We can see that as population density rises, intensifying the struggle for limited resources, sN increases and reduces the reproduction and survival, or the rate of increase, of the average individual. In this way we can think of the individual rate of increase, and through it the growth rate of the population, as being *regulated* by the density of the population. When we introduce this idea of *density-dependent regulation* into our population model we obtain a system composed of two processes: a positive growth process and a negative regulation process (Figure 2.13). The latter is negative because the output variable R is inversely related to the input variable, population density. The interaction between these two processes gives rise to two feedback loops. In addition to the original growth loop (shown as a solid line in Figure 2.13) we now have a negative feedback loop (shown as a broken line). We can see that this loop has a total negative feedback effect because it is composed of one positive and one negative element, the product of which is negative [i.e., $(+)(-) = (-)$]. The equation for this system can be obtained by substituting equation (2.5) for R in equation (2.4) to yield

$$N_t = N_{t-1} + (R_m - sN_{t-1})\, N_{t-1} \qquad (2.6)$$

Although the behavior of this system is determined, in part, by its feedback structure it is also strongly influenced by the parameters R_m and s, which themselves are affected by the quality of the environment and the genetic makeup of the population. To understand the effects of these factors we need to evaluate the dynamics of the model.

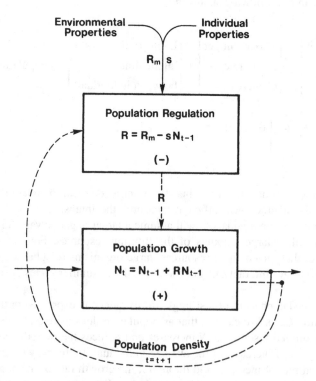

Fig. 2.13 A simple population model governed by positive and negative feedback; the negative feedback loop is shown as a broken line

There is one problem with the equation (2.6). If N_{t-1} is large, then the population number at time t, N_t, predicted by (2.6) may become negative, which does not have any biological meaning. The following modification of the logistic equation is then often used:

$$N_t = N_{t-1} e^{Rm(1 - N_{t-1}/K)} \qquad (2.6a)$$

Because the "regulatory term" $R_m (1 - N_{t-1}/K)$ is now in the exponent, the value $e^{Rm}(1 - N_{t-1}/K)$ is never negative and so is the resulting prediction of N_t. The simulation of equation (2.6a) can be found in the disk that comes with this book, in the sheet "Modified logistic". Alternatively, we can use the original equation (2.6), but instead of the population numbers, we use their logarithms. As the value of logarithm can be negative, the problem of negative predictions of N_t then becomes irrelevant.

2.4. Analysis of the Model

The dynamics of the model can be evaluated by starting at a particular population density, say 10 individuals per unit area of environment, which have a given maximum individual rate of increase, say $R_m = 1$, and a given interaction coefficient, say $s = 0.001$. The population density at the end of the first time increment is then

$$N_1 = N_0 + (R_m - sN_0)N_0$$
$$N_1 = 10 + (1 - 0.001 \times 10) \times 10 = 19.9$$

After another time period the population will be

$$N_2 = 19.9 + (1 - 0.001 \times 19.9) \times 19.9 = 39.4.$$

Continuing this procedure will yield the growth trajectory shown in Figure 2.14. According to this, the population grows rapidly at first but then slows down as it approaches an equilibrium density of 1000 individuals. The equilibrium density, labeled K in Figures 2.12 and 2.14, is a characteristic of the model, which is attained when the per capita rate of increase has declined to zero. It is often referred to as the *carrying capacity of the environment* because it represents the population density where all living space is fully utilized and there is no more room for additional growth. We can see from Figure 2.12 that K is related to both R_m and s because the slope of the line, s, can be expressed as

$$s = R_m / K,$$

so that

$$K = R_m / s.$$

In our example, therefore (Figure 2.14):

$$K = 1 / 0.001 = 1000.$$

The important thing to notice from this analysis is that environmental and genetic inputs, as reflected in R_m and s, determine the level at which the population comes into equilibrium with its environment; that is, they determine its carrying capacity. However, the mechanisms that *control* the growth and equilibrium behavior are contained within the feedback structure. We can also see that larger organisms, with their greater demand for resources and correspondingly larger s values, will have lower carrying capacities than smaller organisms.

Equation (2.6) can also be written in terms of the carrying capacity by substituting R_m/K for s; that is:

$$N_t = N_{t-1} + (R_m - R_m N_{t-1} /K) N_{t-1} \qquad (2.7a)$$

or

$$N_t = N_{t-1} + R_m (1 - N_{t-1} /K) N_{t-1}. \qquad (2.7b)$$

This formulation is similar to the so-called "logistic" equation for population growth, which was first proposed by the mathematician Verhulst in 1839 (the analogous differential equation, which is more commonly encountered in ecological

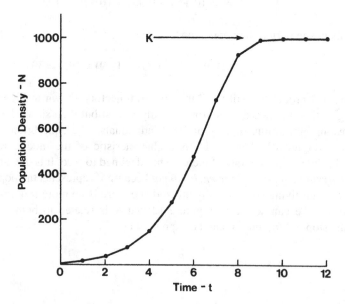

Fig. 2.14 A population growth trajectory computed from the model in Figure 2.13 and equation (2.6), when $N_0 = 10$, $R_m = 1$, and $s = 0.001$, and which equilibrates at a density $K = 1000$

texts, is derived in Note 2.6). The term "logistic" calls attention to the logistical problem of allocating scarce resources to an expanding population. The economics of the system are reflected in the term N_{t-1}/K, which – in effect – represents the demand/supply relationship. For example, if we have U units of an essential resource, and if each organism requires u units to maintain itself, or to replace itself with an offspring should it die, then the maintenance demand of the population is uN, and the demand/supply ratio is $(uN)/U$. From this we can see that the carrying capacity $K = U/u$, or the resource supply divided by the maintenance demand of the individual. Thus, the carrying capacity is defined as the total population that the resources in a given environment can support.

Some natural populations seem to exhibit growth patterns, which are very similar to that shown in Figure 2.14 (e.g., the barnacle population in Figure 2.5). This smooth, or asymptotic, approach to a stable equilibrium should lead us to suspect that the negative feedback mechanisms operate very quickly, or at least very gently, to regulate population growth. However, we also know that, although negative feedback loops tend to create equilibrium conditions, these equilibria are not necessarily stable (Chapter 1). Instability may result when time delays are present in the feedback loops and if the system approaches its equilibrium level too rapidly. Now we can see from equation (2.7) that a time delay is, in fact, present in our model because population density at a particular point in time, t, is determined by its density in the preceding time period, $t - 1$. Thus, the system should become unstable as its rate of approach towards equilibrium gets large, and as this rate depends on the maximum per capita rate of increase, then unstable behavior should occur when R_m becomes large. For example, let us examine the steady-state behavior of the model when R_m is twice that in the first simulation. If we displace the population from its equilibrium density of 1000 by a small number, say 10 individuals, then $N_0 = 1000 - 10 = 990$, and

$$N_1 = N_0 + R_m (1 - N_0/K) N_0$$
$$= 990 + 2 \times (1 - 990/1000) \times 990 = 1009.8$$
$$N_2 = 1009.8 + 2 \times (1 - 1009.8/1000) \times 1009.8 = 990$$

and so on (Figure 2.15). The system seems to be on the verge of instability because the over- and undershoots are of equal size, and there is no tendency for the oscillations to dampen out. The behavior of the model is somewhat reminiscent of the fluctuations seen in the bird populations illustrated in Figures 2.6 and 2.7.

If we continue to perform steady-state analyses we can show that the system is unstable whenever R_m is greater than 2 (Figure 2.16A). However, we can also obtain a general solution for the model's stability in the following way: knowing that the system becomes unstable when the overshoot is larger than the initial displacement, then the criterion for instability is that

$$y / x > 1,$$

where x is the initial displacement and y is the overshoot. We can further show (see Note 2.7) that

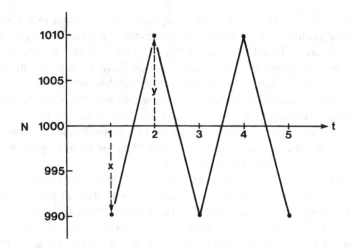

Fig. 2.15 Steady-state behavior of the model specified by Figure 2.13 and equation (2.7) when $R_m = 2$, $K = 1000$, and the initial displacement from equilibrium $x = -10$

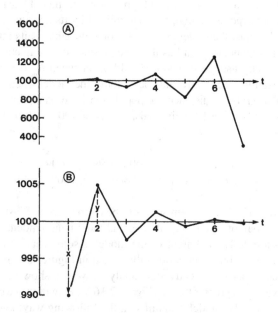

Fig. 2.16 Steady-state behavior of the model when $K = 1000$, $R_m = 3$ (A), and $R_m = 1.5$ (B) and the initial displacement from equilibrium $x = -10$.

$$y / x \approx R_m - 1 \quad (\approx \text{means approximately equal to}) \tag{2.8}$$

which means that the system becomes unstable as soon as the individual rate of increase exceeds 2. Thus, the situation simulated in Figure 2.15, where R_m is exactly 2, is a unique case, which is right on the borderline of instability. It is, in fact, a neutrally stable situation in which the amplitude of oscillation is determined by the magnitude of the initial displacement. For all other values of R_m the system is either stable or unstable, and the steady-state behavior of the model is defined as follows:

1. Unstable oscillations of increasing amplitude when $R_m > 2$ (Figure 2.16A) (see also Note 2.8).
2. Neutrally stable oscillations with amplitude determined by the initial displacement when $R_m = 2$ (Figure 2.15).
3. Damped stable oscillations when $1 < R_m < 2$ (Figure 2.16B).
4. Asymptotic approach to equilibrium when $R_m \leq 1$ (Figure 2.14).

It is possible, of course, for longer time delays to be present in the negative feedback loop. For example, suppose that density-dependent interactions are affected by the size of the population two time increments in the past. The system equation will now be

$$N_t = N_{t-1} + (R_m - sN_{t-2}) N_{t-1}, \qquad (2.9a)$$

or

$$N_t = N_{t-1} + R_m (1 - N_{t-2}/K) N_{t-1}. \qquad (2.9b)$$

Let us evaluate the dynamics of this equation using the same parameter values as we did in Figure 2.14. To start we need to know the population density for the first two time increments. Allowing $N_0 = N_1 = 10$, we can calculate

$$\begin{aligned}
N_2 &= N_1 + (R_m - sN_0) N_1 \\
&= 10 + (1 - 0.001 \times 10) \times 10 = 19.9 \\
N_3 &= 19.9 + (1 - 0.001 \times 10) \times 19.9 = 39.6 \\
N_4 &= 39.6 + (1 - 0.001 \times 19.9) \times 39.6 = 78.4
\end{aligned}$$

and so on (Figure 2.17). We can see that the additional time delay in the negative feedback component causes a system, which previously equilibrated asymptotically, to produce cyclic dynamics. As we would expect, the delay has introduced additional instability. In fact, the stability criterion of equation (2.8) is more correctly written (see Note 2.9)

$$y/x \approx R_m T - 1, \qquad (2.10)$$

where T is the length of the time delay, and stability is a quality of both R_m and T.

The steady-state behavior of the model with a delay of two time periods can be evaluated in the following manner: Suppose we have a system in equilibrium, and whose parameters are $R_m = 1.5$, and $K = 1000$, and we disturb it by removing 10 individuals. The starting population densities will be $N_0 = 1000$, $N_1 = 990$. From this

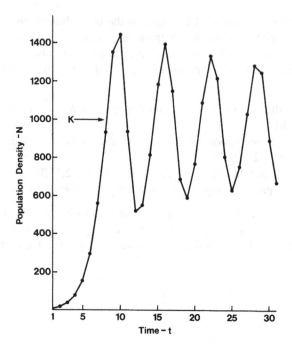

Fig. 2.17 Population growth predicted by equation (2.9) when $N_0 = N_1 = 10$, $R_m = 1$, $s = 0.001$ or $K = 1000$, and the density-dependent response is delayed by two time periods, $T = 2$ - otherwise the model is identical to that produced in Figure 2.14

we can compute $N_2 = 990$, $N_3 = 1004.8$, $N_4 = 1019.9$, using equation (2.9b). You will find that the overshoot is only completely expressed after four time intervals (the student is encouraged to go through these calculations for a number of further increments). The overshoot ratio can now be computed from

$$y / x = (N_4 - K)/10 = 19.9/10 = 1.99.$$

As $y/x > 1$, then the system is unstable. Of course, we could have calculated the approximate overshoot ratio much more easily from equation (2.10); that is:

$$y / x \approx R_m T - 1 = (1.5 \times 2) - 1 = 2.$$

The effect of time delays in the density-dependent feedback processes can be summarized as follows: (1) When there is no delay, then $T = 0$, $R_m T = 0$, $y/x = -1$, and the system will approach equilibrium asymptotically regardless of the value of R_m. This condition is not possible in our discrete-time model because $T > 0$, but it prevails in most continuous-time formulations of the logistic equation (however, see Note 2.10). (2) When the time delay is greater than zero, then stability is determined by the product $R_m T$, and we find instability whenever $R_m T > 2$, stable cycles when $R_m T = 2$, damped-stable oscillations when $1 < R_m T < 2$, and asymptotic stability when $R_m T \leq 1$. However, as shown in Figure 2.17, a delay of two time periods caused the population to cycle smoothly around equilibrium in contrast to the sharp oscillations, which we got with a unit time delay. Thus, long time delays tend to increase the length of the period between cycles as well as the amplitude of displacement during the cycles.

2.5. Environmental and Genetic Effects

We have seen in Figure 2.13 that the properties of the environment and of the individuals making up the population influence the maximum individual rate of increase, R_m, and the interaction coefficient, s. Thus, populations living in different environments, or with different genetic structures may behave quite differently. For example, suppose that a population grows to equilibrium in a particular environment with a given maximum rate of increase, say $R_m = 1.2$ as in Figure 2.18A. Then suppose that the environment becomes more favorable so that R_m increases to 1.8. As we can see in Figure 2.18A, not only has the environmental change raised the equilibrium density, K, but it has also caused the population to be less stable at equilibrium. It may be interesting to compare this simulated population with the experiment in environmental alteration illustrated by Figure 2.7.

It is even more intriguing to consider the effects of environmental or genetic differences on the dynamics of populations that are regulated by lethargic (time delayed) density-dependent processes. In such cases we may find the population going through a series of regular cycles in the more favorable environments, where R_m is large, whilst in less favorable environments the population may remain at relatively constant densities (Figure 2.18B). As we saw earlier, some natural populations seem to exhibit similar behavior in different environments (see also Notes 2.2 and 2.3).

We have now produced an elementary population model, which can produce an array of dynamic behavior depending on the values given its three input parameters. To account for time delays of any length, the model may be written

$$N_t = N_{t-1} + (R_m - sN_{t-T})\, N_{t-1}, \tag{2.11a}$$

or

$$N_t = N_{t-1} + R_m\, (1 - N_{t-T}/K)\, N_{t-1}. \tag{2.11b}$$

This model generates exponential growth trajectories like Figure 2.3 when R_m is positive and the starting population density is very small relative to the equilibrium density K. Conversely, when R_m is negative we get exponential decline to extinction in a similar manner to Figure 2.4. The model also exhibits steady-state behavior at equilibrium, which may be asymptotically stable, as Figure 2.5, or may show sharp oscillations (Figure 2.6) or cycles (Figure 2.9) depending on the magnitude of the

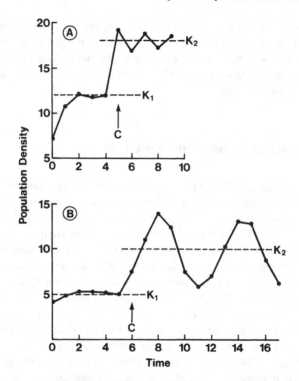

Fig. 2.18 Dynamics of populations governed by equation (2.11) when subjected to environmental changes at a particular time, C, which affect the maximum per capita rate of increase, R_m: (A) R_m = 1.2 prior to time C and 1.8 afterwards, $s = 0.1$, $T = 1$: (B) R_m = 0.5 prior to time C and 1.0 afterwards, $s = 0.1$ $T = 2$

time delay and the maximum per capita rate of increase. We can also visualize how changes in the environment, acting through the maximum individual rate of increase, can cause changes in the equilibrium density (Figure 2.7) and suppress or induce cyclic behavior. These are encouraging results, which give us some confidence in the structural soundness of the model. However, there are still some conceptual weaknesses. In particular, the density-dependent feedback structure remains rather mysterious and retains the weak assumption of linearity. There are also problems in interpreting the eruptive kinds of behavior shown in Figure 2.8. To sharpen our concept of the population system we will explore the mechanisms of density-dependent population regulation in more detail in the next chapter.

2.6. Chapter Summary

In this chapter we have defined a population system, looked at some of the dynamic patterns that natural populations exhibit, and have constructed and analyzed an elementary population model. The main points are summarized below:

1. A population system consists of a number of interacting or intercommunicating individuals of the same species, which coexist within certain geographic boundaries.
2. The environment provides the population system with inputs such as food, nesting sites, space to hide from or escape predators, parasites, diseases, and competitors, and may also supply immigrants into the population. The environment may receive outputs from the population system in the form of depleted resources, pollution, and emigrants.
3. We also consider the basic physiological and behavioral properties of the individuals making up the population to be inputs into the system. These qualities, acting in conjunction with the environment, govern the processes of natality, mortality, and migration, which control the state of the system.
4. Natural population systems seem to exhibit four basic patterns of behavior: (a) Exponential, or geometric, growth and decline, depending on the favorability of the environment, which is governed by a positive feedback loop. (b) Steady-state behavior about an equilibrium density, which is controlled by a negative feedback loop. The equilibrium density, or reference level, is set by environmental and individual properties, and the steady-state behavior may be characterized by gentle or violent oscillations around equilibrium. (c) Cycles of a 4- to 5- or 8- to 10-year period caused by time delays in the negative feedback loop. These cycles may be synchronized over broad geographic regions, probably by environmental disturbances, and their amplitude is strongly influenced by environmental conditions, to the extent that they may be completely suppressed in unfavorable environments. (d) Erratic population fluctuations may be exhibited by populations inhabiting extremely variable environments.
It should be emphasized that a particular population may exhibit any of these basic patterns over a specific time period, and may switch from one to another as environmental conditions change.
5. The positive feedback loop was described in the equation

$$N_t = N_{t-1} + RN_{t-1},$$

where N_t is the density of the population at time t, and R is the per capita rate of increase as determined by the processes of natality, mortality, and migration.
6. Density-dependent negative feedback was expressed by an inverse linear relationship between the individual rate of increase and population density

$$R = R_m - sN_{t-T},$$

where R_m is the maximum per capita rate of increase, s represents the inhibitory effect of each individual on the rate of increase of its cohorts, and T is the time delay in the negative feedback response.
7. The stability of the negative feedback loop is determined by the maximum individual rate of increase, R_m, and the length of the delay in the feedback response,

T, such that the loop is unstable when $R_m T > 2$, neutrally stable when $R_m T = 2$, damped-stable when $1 < R_m T < 2$, and asymptotically stable when $R_m T \leq 1$.

8. Environmental and genetic changes, acting through the maximum individual rate of increase, R_m, or the interaction coefficient, s, may cause dramatic changes in the dynamic behavior of the population system.

Exercises

2.1. Suppose we have a population of 100 individuals and we observe that 20 new individuals are born during the following year, 10 die, 3 immigrate, and 5 emigrate in the same time period.

   A. Calculate the per capita rate of increase, R.
   B. Predict population density for the next 15 years, assuming that the per capita rate of increase remains the same over this time. Use the disk that comes with this book for the simulation.
   C. What is unreasonable about this prediction?
   D. What will happen to the population if 15 individuals die per year, 3 immigrate and 8 emigrate, assuming that births remain the same? Calculate the per capita rate of increase, R and then use the disk that comes with this book for the simulations.

2.2. A population is observed to remain at a relatively constant density of about 2000 for many years.

   A. What processes may be involved in maintaining this *status quo*?
   B. Suppose an environmental catastrophe reduced this population to 200 individuals and, after the catastrophe, we observed that 400 new individuals were born, 100 immigrated, 20 emigrated, and 180 died during the following year. Calculate the per capita rate of increase in the year after the catastrophe.
   C. Calculate the maximum per capita rate of increase under the assumption that density-dependent processes act linearly and that the carrying capacity remains the same as before the catastrophe.
   D. Calculate the density-dependent coefficient, which represents the inhibitory effect of each individual on its cohorts.
   E. Using the disk that comes with this book, plot the trajectory this population will take over the next 15 years, assuming that the environment remains consistently favorable during this time.
   F. Describe and explain the equilibrium behavior and stability properties of this population.

2.3. Evaluate the steady-state behavior of the model

$$N_t = N_{t-1} + R_m(1 - N_{t-T}/K) N_{t-1}$$

by plotting the dynamics for six time periods following an initial displacement of
−10 from equilibrium, when the parameters are

    A. $R_m = 0.8$, $K = 1000$, $T = 1$;
    B. $R_m = 0.8$, $K = 10,000$, $T = 1$;
    C. $R_m = 0.8$, $K = 100,000$, $T = 1$;
    D. $R_m = 1.8$, $K = 1000$, $T = 1$;
    E. $R_m = 2.8$, $K = 1000$, $T = 1$;
    F. $R_m = 0.8$, $K = 1000$, $T = 2$;
    G. $R_m = 0.8$, $K = 1000$, $T = 3$ (calculate for 200 time intervals).

Use the disk that comes with this book for the simulation. Calculate the overshoot
ratio, y/x, from the graphs you make and check your answers against the equation

$$y / x \approx R_m T - 1.$$

What neighborhood stability qualities does the model have under the conditions A
through G?

Notes

2.1. The definition of population given in this book is purposely loose to allow
flexibility in defining the geographic bounds of particular populations. As
such it is distinct from the more rigid views of the taxonomist who, because
he deals with the evolution of species from geographically isolated populations, insists that populations be separate from, and not interbreed with, other
similar populations. In this strict definition, population boundaries are determined by barriers to migration, rather than arbitrarily determined boundaries.

2.2. Much has been written concerning the causes of cycles in small mammal populations. Theories ranging from sunspots to physiological and genetic selection have been erected and argued over. Although we have restricted ourselves
to inferences gleaned from general systems theory alone, some references are
included for those who may be intrigued by this subject.

    Wildlife's Ten-Year Cycle by L. B. Keith, published by the University of
    Wisconsin Press, Madison, 1963.

    A paper by C. J. Krebs in *Population Ecology*, edited by L. Adams, published
    by Dickerson Publishing Co., Inc., Belmont. California, 1970, gives a
    review of current theories for the causes of lemming cycles.

    Animal Population Ecology by J. P. Dempster, published by Academic Press,
    Inc., London, 1975, provides a number of examples of cyclic animal
    populations.

2.3. The effect of elevation on the cyclic behavior of larch budmoth populations can
be seen in Werner Baltensweiler's paper in *Dynamics of Numbers in Populations*,

printed by the Centre for Agricultural Publishing and Documentation, Wageningen, Netherlands, 1971. Similar phenomena, where cycles occur in certain environments and not in others, have been noticed with other forest insects (e.g., A. A. Berryman, *Canadian Entomologist*, vol. 110, p. 513, 1978).

2.4. Equation (2.4) can be written as

$$N_t - N_{t-1} = RN_{t-1} = \Delta N,$$

where ΔN represents the change in N over the time increment $t - 1$ to t. To obtain the instantaneous rate, we divide through by the time interval Δt to give

$$\frac{\Delta N}{\Delta t} = \frac{RN}{\Delta t}$$

If we let $R/\Delta t = r$, the instantaneous rate of increase, and allow the time interval to become very small, then the equation can be written in continuous time

$$\frac{dN}{dt} = rN,$$

where dN/dt represents the change in population density in an instant of time. This equation can be solved to yield the continuous time equation

$$N_t = N_0 \exp(rt).$$

where N_0 is the initial density. From this we can compute population density after any length of time in a single step, whereas our discrete-time model had to be solved one step at a time. For example, when computing curve A in Figure 2.11 we calculated N_t for four time intervals to arrive at the density 50.62. Using the continuous time equation with $r = 0.41$ gives

$$N_4 = 10 \text{ x } \exp(0.41 \text{ x } 4) = 50.625.$$

In order to do this we have to obtain equivalence between R and r. This is done by setting the time increment to unity, so that

$$N_t = N_{t-1} \exp(r)$$
$$N_t = N_{t-1} = N_{t-1}[\exp(r) - 1]$$

and

$$\frac{N_t - N_{t-1}}{N_{t-1}} = \exp(r) - 1.$$

From text equation (2.3) we see that

$$R = \frac{N_t - N_{t-1}}{N_{t-1}}$$

and therefore

$$R = \exp(r) - 1,$$

or

$$r = \log_e(R+1).$$

Although continuous-time equations are much more elegant and amenable to sophisticated mathematical analysis, they become very difficult to solve in complex systems. Although the discrete-time equation has to be solved by repeated calculation, its structure is readily apparent and, as we shall see later, this transparent structure will be helpful in our attempts to understand more complicated systems. For this reason discrete-time models will be used throughout the text, although their continuous analogues will be given in the notes when appropriate and possible. You will also find many examples of continuous systems in the disk that comes with this book.

2.5. For those students interested in the historical development of ecology as a science we recommend the book *Principles of Animal Ecology* by W. C. Allee, A. E. Emerson, O. Park, and T. Park, published by W. B. Saunders Co., Philadelphia, 1949. These authors note that Machiavelli and Giovanni Botero both anticipated Malthus' ideas over 200 years before his book was published.

2.6. We can rewrite equation (2.7) (page 44) as

$$N_t = N_{t-1} + R_m (1 - N_{t-1}/K) N_{t-1}$$

or

$$\frac{\Delta N}{\Delta t} = \frac{R_m}{\Delta t}(1 - N_{t-1}/K) N_{t-1}$$

and, if we allow the time increment Δt to become infinitesimally small then, as $R_m/\Delta t = r_m$, we get

$$\frac{dN}{dt} = r_m(1 - N/K) N,$$

which is the familiar "logistic" equation, first published in 1838 by P.F. Verhulst in his paper Recherches mathematiques sur la loi d'accrossement de la population in Memoirs de l'Academie Royal Bruxelles, pp. 1–38. The instantaneous per capita rate of increase, r_m, is the maximum possible rate of increase in a given environment. This equation is stable under all conditions because the time delay is effectively zero. As we found in Note 2.4

$$r_m = \log_e(R_m + 1).$$

2.7. Proof that the overshoot ratio $y/x \approx R_m - 1$:

Suppose we have a population at equilibrium, K, and we displace it by an extremely small amount, say $-x$, so that $N_1 = K - x$. Then, from equation (2.7) (page 44):

$$N_2 = N_1 = R_m(1 - N_1/K)N_1$$

and substituting $K - x$ for N_1 we get

$$\begin{aligned} N_2 &= K - x + R_m[1 - (K - x)/K](K - x) \\ &= K - x + R_m(x/K)(K - x) \\ &= K - x + R_m x(1 - x/K) \end{aligned}$$

Now as the overshoot of the equilibrium position is $y = N_2 - K$, then

$$y = R_m x(1 - x/K) - x$$

and the overshoot ratio becomes

$$y/x = R_m(1 - x/K) - 1.$$

The initial disturbance, x, was extremely small relative to K and so we can assume that $x/K \approx 0$. From this it follows that

$$y/x \approx R_m - 1.$$

Note that by making the assumption that x is a very small displacement, we are restricting our stability analysis to the immediate vicinity of K. The *neighborhood stability* of a system is its stability close to the equilibrium point and is distinct from its *global stability*, where disturbances of any magnitude must be considered (see Chapter 1). In many systems, including the model we are analyzing, neighborhood and global stability are equivalent. However, more realistic biological models, as we shall see later, are frequently nonlinear and, in such cases, a neighborhood analysis may not define the system's global stability.

2.8. Robert M. May [*Science (Washington)*, vol. 186, p. 645, 1974] has shown that, as R_m becomes larger than 2, a rather surprising array of dynamic behavior emerges from this simple model. When $2 < R_m < 2.57$ the unstable oscillations settle down into stable limit cycles of period 2^n, where n is the number of points in the cycle; for example, when $2 < R_m < 2.449$, we get a 2-point stable cycle (period 2^1). However, as R_m increases above 2.449 this cycle becomes unstable but then settles into a $2^2 = 4$-point stable cycle, and so on. Check that these values are correct using the disk that comes with this book. Those interested in the mathematics of this phenomenon are referred to a paper by R. M. May and G. F. Oster in the *American Naturalist*, vol. 110, p. 573, 1976.

2.9. Proof that the overshoot ratio $y/x \approx R_m T - 1$:

Suppose we have a system that is described by the equation

$$N_t = N_{t-1} + R_m (1 - N_{t-2}/K) N_{t-1},$$

which specifies that negative feedback acts with a delay of two time increments. Now if we disturb this system from its equilibrium at K by a very small amount, say $-x$, then

$$N_1 = K - x.$$

After the next time period the system will move to

$$N_2 = N_1 + R_m (1 - N_0/K)N_1$$

but, as $N_0 = K$, then

$$N_2 = N_1 + R_m (1 - K/K)N_1 = N_1.$$

As we cannot observe an overshoot in this first time period we must continue:

$$N_3 = N_2 + R_m (1 - N_1/K) = N_2.$$

However, as $N_1 = N_2$ this equation becomes identical to that for a time delay of only one period. To obtain an overshoot resulting from a delay of two periods we must compute the dynamics over a further time increment:

$$N_4 = N_3 + R_m (1 - N_2/K) = N_3.$$

Substituting the previous equation for N_3 we get

$$N_4 = N_2 + R_m (1 - N_1/K) N_2 + R_m (1 - N_2/K) [N_2 + R_m (1 - N_1/K) N_2]$$

and as $N_2 = N_1 = K - x$, and simplifying as we did in Note 2.7, we get

$$N_4 = K - x + R_m x(1 - x/K) + R_m x(1 - x/K) + R_m (x/K) [R_m x(1 - x/K)].$$

With our assumption that x is an extremely small displacement relative to K so that $x/K \approx 0$, then

$$N_4 \approx K - x + R_m x + R_m x + 0$$

and as the overshoot $y = N_4 - K$, then the overshoot ratio is

$$y / x \approx 2R_m - 1.$$

We can perform the same, though much more laborious, analysis with larger time delays and show that, in general

$$y / x \approx R_m T - 1.$$

Remember from Note 2.7 that this is a neighborhood analysis, which only defines the global stability of linear systems.

2.10. The influence of time delays on the behavior of the continuous "logistic" population model was first described by G. E. Hutchinson in the *Annals of the New York Academy of Science*, vol. 50, p. 221, 1948. For those interested in a more rigorous discussion, the book *Stability and Complexity of Model Ecosystems* by R. M. May, Princeton University Press, 1975, is recommended.

2.11. For those interested in a more thorough, but user-friendly analysis of continuous systems we recommend the book *A Primer of Ecology* by N. J. Gotelli, published by Sinauer Associates, Inc., Sunderland, Massachusetts, 1998. A highly advanced analysis of population systems, including stochastic effects, is given in the book *Analytical Population Dynamics* by T. Royama, published by Chapman & Hall, London, 1992. Its full understanding, however, requires a great deal of mathematical knowledge.

Chapter 3
Population Regulation and a General Model

In the last chapter we discussed the general concept of density-dependent negative feedback and its effect on the individual rate of increase. As we were most interested in the basic feedback structure of the system, we did not concern ourselves unduly with the mechanisms involved, or with a correct form for the negative feedback function. In this chapter, therefore, we will look at the biological processes involved in the regulation of population growth with an eye to developing a more general model of the population system.

3.1. Density-Dependent Mechanisms

Most ecologists now accept the proposition that the growth rate of a population must be related to its density. However, this was not always the case and, in the past, there has been considerable debate over the relative importance of different regulating mechanisms (see Note 3.1). Biologists working with small organisms inhabiting harsh physical environments frequently concluded that population density was controlled by the physical properties of the environment. On the other hand, those engaged in research in more benign environments, or with larger organisms, which are less affected by their physical environment, often concluded that populations were regulated by density-dependent negative feedback. Nowadays, however, most population ecologists accept the comprehensive view that density is regulated by a complex of factors pertaining to the population system and its environment *as a whole*, although at any particular time one or several of these factors may be playing a decisive role in limiting population growth. We came to this same basic conclusion in Chapter 2 when we recognized the importance of the environment in setting the reference level for density-dependent regulation. Obviously, when the environment is changing constantly, populations will be continuously growing (favorable environment) or declining (unfavorable environment), giving the impression that the environment alone is controlling population size. In more consistent environments, however, population densities tend to remain relatively constant, suggestive of strong negative feedback control.

A.A. Berryman, P. Kindlmann, *Population Systems: A General Introduction*
© Springer Science+Business Media B.V. 2008

3.1.1. Competitive Processes

As both Malthus and Darwin realized, the most likely mechanism for the regulation of population density is the spontaneous competition that occurs amongst crowded individuals for limited environmental resources. The effects of this struggle for existence may be manifested in many ways: Some individuals may not obtain enough food to support life and so die of starvation. Others may survive, but with their reproductive capabilities reduced because of improper nourishment. Weakened individuals may be more vulnerable to predators and parasites and less resistant to diseases. In addition to food resources, organisms may also compete for space in the environment and, under crowded conditions, some individuals may be unable to find nesting sites or hiding places from their predators and parasites. Crowding may also cause subtle changes in the normal patterns of individual behavior, which may result in increased emigration out of the crowded regions and, in extreme cases, cannibalism and aberrant sexual behavior (see Note 3.2). The sum of all these effects of competition produces higher death and emigration rates, and lower birth and immigration rates, as the density of the population rises (Figure 3.1).

Some organisms possess behavioral mechanisms that help them to avoid the wasteful scramble for resources, which often leads to everyone getting something but nobody receiving enough to survive and reproduce. Territorial behavior ensures that those who win a territory obtain sufficient resources while the losers are left to

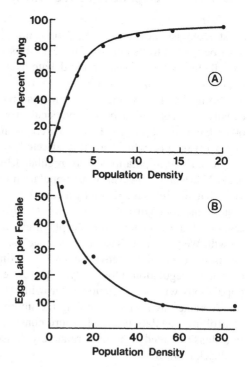

Fig. 3.1 The effects of population density on (A) mortality of a beetle feeding on grain [redrawn from A. C. Crombie, *Proceedings of the Royal Society (Section B)*, vol. 131, p. 135, 1944], and (B) the eggs laid by plant-feeding bugs (redrawn after L. R. Clark, *Australian Journal of Zoology*, vol. 11, p. 190, 1963)

fend as best they can. This may be a more efficient way of allocating scarce resources, but the essential ingredients of competition are retained in the struggle to obtain and defend a territory.

The effects of competition for scarce resources on the reproduction, survival, and migration of individuals are usually manifested quite rapidly so that time delays in the negative feedback loop are relatively short. Therefore, competitive interactions should stabilize populations at a characteristic equilibrium density set by the level of environmental resources. In other words, the reference level for the negative feedback is determined by environmental properties (see Figure 2.18). However, if the population has a negative effect on these environmental properties, then the reproduction and survival of future generations may also be affected. For example, a population may consume its food supply faster than it can be regenerated, in which case future populations will suffer a shortage of this resource. The accumulation of waste materials, or pollutants, will have similar effects because they tend to have a greater impact on future populations than on those, which produced them. Delayed feedback can also occur through the response of predators, parasites, and diseases that are present in the environment. Predatory species often migrate into areas where their prey are abundant, or their numbers may increase because a plentiful food supply means greater reproduction and survival. However, as it takes time for predators to locate dense prey populations, and to convert their food into offspring, their numerical response to prey density will be delayed somewhat. The length of this time delay will depend on their efficiency at locating prey concentrations and on their fecundity and frequency of reproduction.

In general terms this means that whenever a population influences the properties of its environment, either through pollution, overexploitation of resources, or encouraging the buildup of natural enemies, then the effects will usually be transmitted, with a time delay, to future populations. As we have seen in Chapters 1 and 2, delayed feedback can give rise to population cycles, such as those exhibited by the snowshoe hare and its predator, the Canadian lynx (Figure 3.2). These cycles could be caused by delays in the response of the lynx to the hare population, or the hare to its food supply or, more likely, a combination of both. Whatever the specific cause, our general rule states that the hare or lynx populations must affect, in some

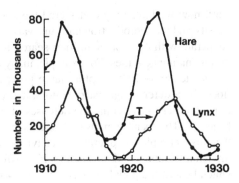

Fig. 3.2 Cycles in the population dynamics of the snowshoe hare and its predator, the Canadian lynx. T indicates approximate time delay in response of lynx to hare population change (Redrawn after D. A. MacLulich, *University of Toronto Series in Biology*, no. 43, p. 5, 1937.)

way, the properties of their own environments in order to create delayed feedback and the resultant population cycles.

Delayed feedback may also occur through the effect of population density on its genetic properties. For instance, we might expect the weaker genotypes to succumb first to the effects of intense competition. They may be more vulnerable to predation or disease, or less capable of grasping the disputed resources and, therefore, die of starvation. In addition, the stronger, more vigorous genotypes would tend to move out of the crowded regions in search of "greener pastures" (this is covered in more detail in Chapter 5). If these genotypes have different reproduction and survival characteristics, as we would expect, then these effects will be transmitted to future generations. Thus, genetic feedback can also create time delays, which may give rise to population cycles (see Note 3.3).

In certain cases genetic changes may become relatively permanent, giving rise to evolutionary trends. We usually think of evolution as the long-term formation of new species through the processes of mutation and natural selection. Although the study of population dynamics usually involves much shorter time periods, so that mutations can normally be ignored, we must be concerned with the *adaptation* of populations to their environments and to their own densities. For example, individuals with an exceptional ability to escape predators may be selected for so that their genotype makes up an increasingly large part of the population, and also alters the population growth rate. The population may then grow for a time but the increasing density will make prey capture easier for the predators and they will again exert their effect. In addition, the predators may also be subjected to evolutionary pressure because of the difficulty they experience in capturing prey. This may lead to the *co-evolution* of predator genotypes with greater abilities for hunting and capturing prey. Thus, we may see a continuous genetic jockeying amongst predator and prey genotypes, and the time delays intrinsic to adaptive evolution may cause population cycles (see Note 3.4). Similar co-evolutionary tendencies may also be visualized between competing species. These relationships, and those of predators and their prey, will be explored in more detail later in this book.

3.1.2. Cooperative Processes

Until now we have only considered the negative interactions between population density and the reproduction and survival of individuals. However, organisms often cooperate with each other in their search for food, to escape from predators, and during mating activities. For example, many predators form hunting groups (prides, packs, etc.) in order to capture large prey; fish and birds often form schools and flocks as a defense against predators, and certain insects aggregate their populations in order to overcome the defenses of their host plants (e.g., bark beetles; see Note 3.5). The social animals such as ants, bees, termites, and humans have developed the most complex cooperative behaviors, which may include specialized roles (division of labor) and altruism (self-sacrifice for the good of the group), both of which benefit the population as a whole (Note 3.6).

Cooperative processes have a positive feedback effect because they provide the average individual with a *greater* chance to survive and reproduce as population density *rises*. Thus we see that the survival of bark beetles improves as their density increases because the defense secretions of their coniferous hosts are diluted amongst a large number of individuals (Figure 3.3A), and flour moth females lay more eggs because they have a greater chance of finding a mate as population density rises (Figure 3.3B).

Cooperation between individuals is necessary for mating. Therefore, positive feedback often operates at low population levels because an increase in density provides a greater opportunity for finding mates. This also means that very sparse populations may be in danger of extinction because of difficulties that individuals may have in locating mates. This problem is particularly acute for those species that migrate to distant mating grounds or that have social mating habits. An example is the passenger pigeon, which seems to have become extinct when their populations, being decimated by hunters, became too sparse to effectively maintain their colonial mating habits.

We have now seen that cooperative processes can result in a positive relationship between population density and the reproduction and survival of individual organisms. We would expect this positive feedback effect to be most prominent in

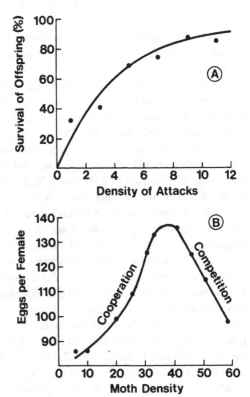

Fig. 3.3 (A) Effect of bark beetle attack density on the survival of its offspring from the defense secretions of its coniferous host (redrawn after A. A. Berryman. *Environmental Entomology*, vol. 3, p. 579. 1974). (B) Effect of population density on the number of eggs laid by flour moth females (redrawn from G. C. Ullyett, *Journal of the Entomological Society of South Africa*, vol. 8, p. 53, 1945)

the lower density ranges and that, as populations become denser, the effects of competition should dominate to create an overall negative feedback loop. Thus, the combination of positive feedback cooperation and negative feedback competition often produces a unimodal relationship between population density and the individual rate of increase (see Figure 3.3B, for example).

3.2. Feedback Integration

Having examined the mechanisms that can be involved in the density-dependent feedback loop, it is now time to see how they operate together to control population growth. When we built the population model in Chapter 2, we made the tenuous assumption that the individual rate of increase was linearly related to population density. However, we have seen that some density-dependent interactions are decidedly nonlinear (e.g., Figures 3.1 and 3.3). The integrated effect of density-dependent feedback can be seen by plotting the individual rate of increase, R, against the density of the initial population, N_{t-1}. The rate of increase is computed from time-series population data by

$$R = (N_t - N_{t-1})/N_{t-1} \qquad (2.3)$$

This is then plotted on N_{t-1}, as is shown in Figure 3.4 for three different sets of data. As we can see, none of these data produce a linear density-dependent relationship.

When populations exhibit cyclic behavior it is impossible to identify density-dependent relationships by plotting R on N_{t-1}, for we will obtain cyclic trajectories (Figure 3.5A). However, we can sometimes find the density-dependent relationship and the magnitude of the time delay by plotting R on N_{t-T}, increasing the delay T until the circular trajectory disappears. With the snowshoe hare data of Figure 3.5, a cyclic pattern is still evident when we plot R on N_{t-2}, but it disappears when the time delay is increased to 3 (Figure 3.5B,C). From this we can infer that negative feedback acts on the hare population with a delay of about 3 years.

In the examples we have looked at so far, the density dependent interaction produces a single equilibrium point, K, where the curve intercepts the $R = 0$ abscissa (see Figures 3.4 and 3.5). However, certain population systems appear to have more than one possible equilibrium position, even under identical environmental conditions. An example of this is the strange story of the odd-year pink salmon run on the Atnarko River in British Columbia (Figure 3.6). During the census period from 1951 to 1965 the population of fish returning to the river to spawn cycled between five hundred thousand and three million individuals. The cyclic pattern indicated that time-delayed feedback mechanisms were involved in regulating the population. In the year 1967, however, the population, which was at the low ebb of its cycle, was drastically reduced by a combination of over-fishing and bad weather. To the surprise of the fishery managers, the run did not recover from this

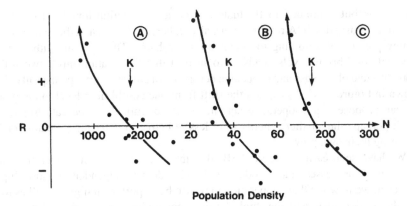

Fig. 3.4 Some observed relationships between the individual rate of increase, R, and population density, N, for (A) the edible cockle (redrawn from D. A. Handcock in the book *Dynamics of Numbers in Populations*, p. 419; see Figure 2.6 for complete reference), (B) the great tit (redrawn from H. N. Kluyver, p. 507 in the same book), and (C) the southern cowpea weevil (redrawn after S. Utida, *Researches on Population Ecology*, vol. 9. p. 1, 1967). Note that the equilibrium population density, or carrying capacity, is indicated by K

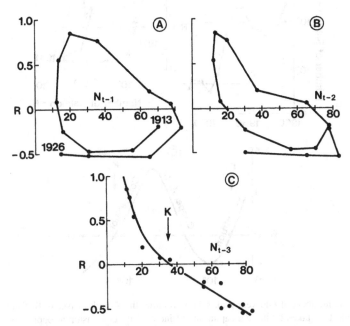

Fig. 3.5 Relationship between the per capita rate of increase, R, and population density, N, for the snowshoe hare 1 (A), 2 (B), 3 (C) years in the past (see Figure 3.2 for reference). Note the equilibrium density for hares, $K \approx 36$

catastrophe but continued to fluctuate around a new equilibrium level of about fifty-five thousand fish. A return to normal weather conditions and the reduction of fishing pressure did nothing to alleviate this problem. This system, with its two apparent equilibrium levels, leads us to suspect that the relationship between the individual rate of increase and population density has a complex form, perhaps like that shown in Figure 3.6C. To explain the shift from one equilibrium level to the other we can propose that cooperative activities were disrupted by the catastrophe of 1967; perhaps the smaller schools were less effective in deterring predators or in capturing their own prey.

We have seen earlier (Figure 3.3B) that interactions between cooperative and competitive processes may produce unimodal density-dependent relationships. However, we now see that cooperation may also be important at high as well as low population densities, and that high-density cooperation may produce even more complicated multimodal curves. Perhaps a clearer picture of this phenomenon is to be seen in certain bark beetles that can only attack very weak trees when their populations are small but can kill relatively healthy trees at high population densities.

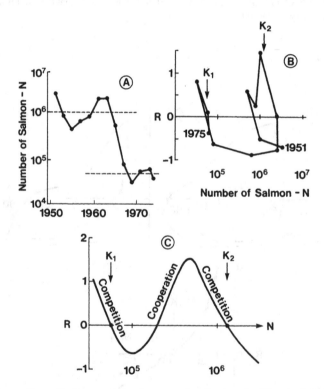

Fig. 3.6 The numbers of odd-year pink salmon running the Atnarko River in British Columbia (A), the trajectory taken by the per capita rate of increase in relationship to population size (B), and the hypothesized interaction between cooperative and competitive processes in determining this relationship (C) (drawn from data in R. M. Peterman, *Journal of the Fisheries Research Board of Canada*, vol. 34, p. 1130, 1977)

This is because the large populations are able to cooperatively overwhelm the defenses of quite vigorous trees. Thus, the interaction between these bark beetles and their hosts may also produce bimodal density-dependent curves of the type shown in Figure 3.6C (see also Note 3.5).

Population systems that exhibit divergent behavior because of multiple equilibrium levels seem to be quite common in nature. Other examples will be examined later in this book and we will look at the mechanisms responsible for maintaining these equilibria in much more detail (Chapter 4). For the present, however, we will leave this interesting topic and return to our modeling exercise.

3.3. A General Population Model

The elementary model we constructed in Chapter 2 performed quite well at simulating the dynamic behavior that was observed in certain real population systems, but we have since uncovered some serious deficiencies. In particular, a general model should consider nonlinear density-dependent processes, cooperative as well as competitive interactions, and delayed feedback operating through the environment or the gene pool of the population.

Let us start from our basic equation for population growth; that is

$$N_t = N_{t-1} + RN_{t-1}, \tag{3.1}$$

where R is the per capita rate of increase in the time interval $t-1$ to t for a population with fixed genetic structure living in a constant environment. When this rate of increase is linearly related to population density at the beginning of the time interval, we can write

$$R = R_0 - sN_{t-1}, \tag{3.2}$$

where R_0 is the limiting condition on R as N_{t-1} approaches zero. In this linear model, of course, $R_0 = R_m$, the maximum per capita rate of increase, but in some nonlinear cases R_0 may not be the maximum (e.g., R_0 is negative in Figure 3.7B,C). Now in this equation the negative sign of s implies that competitive interactions dominate the system over all population densities. As we have seen, however, cooperative interactions may sometimes dominate over certain ranges of population densities. When this occurs the sign of s will change to positive. In other words, the relative dominance of cooperative and competitive interactions may change as population density changes and this will be reflected by the magnitude of the coefficient s. If we assume that s is positive at low density and decreases linearly with increasing population density, then

$$s = s_p - s_m N_{t-1}, \tag{3.3}$$

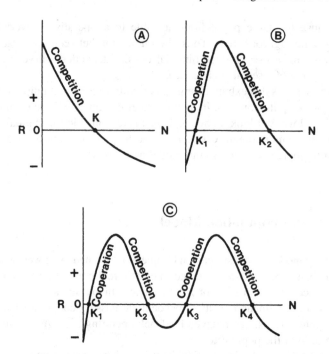

Fig. 3.7 Some possible relationships between the individual rate of increase, R, and population density, N; (A) a curvilinear competitive interaction, (B) cooperation acting at low density and competition at high density, (C) cooperation acting at low and intermediate densities and competition at intermediate and high densities. The K's represent potential equilibrium points

where s_p is the maximum benefit received from cooperative interactions and s_m is the competitive effect which becomes more and more dominant as population density rises. Introducing this expression into equation (3.2) yields

$$R = R_0 + (s_p - s_m N_{t-1}) N_{t-1},$$

or

$$R = R_0 + s_p N_{t-1} - s_m N_{t-1}^2. \qquad (3.4)$$

This second order quadratic equation produces a unimodal individual rate of increase curve of the general form shown in Figure 3.7B. Of course, we can also derive higher order equations to describe the more complex curves (Figure 3.7C). However, our theory will still be constrained by underlying assumptions concerning the form of the cooperative and competitive interactions. To free ourselves from these constraints let us consider R to be an unspecified function of population density

$$R = f(N_{t-1}), \qquad (3.5)$$

which may result in any of the forms shown in Figure 3.7 or in a modification of one of these forms. The only constraint on this general equation is that competitive interactions must eventually dominate to create an upper equilibrium or carrying capacity. However, below this there may be one or more additional equilibrium points created when the relative dominance of cooperative and competitive interactions changes with respect to population density. We will examine the effects of these different kinds of equilibria later in this chapter, but for now we will restrict the discussion to simple competitive systems such as that illustrated by Figure 3.7A.

In our analysis of population systems governed by linear density-dependent relationships (Chapter 2) we came to the conclusion that environmental and genetic properties influenced the maximum individual rate of increase of a population and its density at equilibrium. In other words, we would expect greater rates of increase and higher equilibrium densities in more favorable environments because there will be more food, fewer predators, and so on. Therefore, we can argue that environmental favorability will affect the amplitude, or height, of the basic density-dependent relationship and, through this, it will also affect the equilibrium density (see also Note 3.7). When we introduce environmental favorability as a variable in our model, we obtain a three-dimensional relationship between the individual rate of increase, population density, and environmental favorability as illustrated in Figure 3.8. Note that both the height of the curve as well as its interception with the zero growth plane ($R = 0$) change in direct relationship with the favorability of the environment.

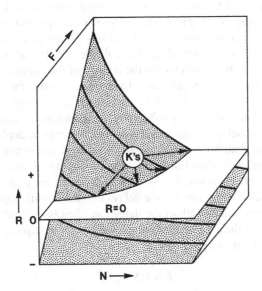

Fig. 3.8 Three-dimensional representation of the interaction between the individual rate of increase, R, population density, N, and environmental favorability, F, where K represents the equilibrium density line at the interception of the curve with the $R = 0$ plane

If we assume that the environment acts as a simple multiplier to the basic density-dependent function, then we can rewrite equation (3.5) as

$$R = f(N_{t-1})F, \tag{3.6}$$

where F is a measure of the relative favorability of the environment over the interval $t-1$ to t. This will be determined by the relative abundance of food, nesting and hiding places, predators, parasites and diseases, as well as climatic and other factors.

The effect of genetic variations may be visualized in a similar manner. Changes in the population gene pool, which affect the reproduction and survival of individuals, will alter the amplitude of the basic density-dependent relationship. Thus, we can incorporate genetic properties, symbolized by G, into the model to yield

$$R = f(N_{t-1})FG, \tag{3.7}$$

if the multiplicative assumption is again made. We should also note that genetic evolution can also modify the shape of the density-dependent relationship because it can alter the basic processes of cooperation and competition. That is, genetic adaptations may lead to different cooperative or competitive strategies, which will change the overall shape of the function. However, as these changes normally take place over rather long time periods they will be ignored for the present.

In equation (3.7) the properties of the environment and the genetic structure of the population act as density-independent inputs into the system. However, if population density influences its environment or gene pool, then these properties become components of feedback loops which may introduce time delays into the density-dependent response. For example, suppose that a population in one time period removes a part of its food resources and that this cannot be replaced by the time it is needed in the next time period. In effect, the favorability of the environment at one point in time has been affected by the density of the population in a previous time period. Whether or not population effects are transmitted through the environment depends largely on the rates of resource depletion and regeneration. That is, environmental favorability is only reduced if the resources are used up faster than they can renew themselves. These processes of environmental depletion and regeneration, and for that matter genetic adaptation as well, are extremely complex in their own right, and we cannot introduce them into our model without greatly complicating the picture. However, we do know that one of the main effects of these feedback processes is to introduce time delays into the density-dependent relationship, and that the average length of the delay can often be inferred from field data (see Figure 3.5). For the present, therefore, let us simply incorporate a variable time delay, T, into our model to give

$$R = f(N_{t-T})FG \tag{3.8}$$

where T may change in accordance with the effect of the population on its environment or genetic structure (Note 3.8).

We now have a general, though still highly simplified, conceptual model of a population system, the feedback structure of which is illustrated in Figure 3.9.
It has four potential feedback loops: (1) a positive growth or decay loop, $A \to B$; (2) a population regulation loop, $A \to C \to D$; which may be positive or negative depending on the relative dominance of cooperative or competitive interactions; (3) an environmental feedback loop, $A \to F \to T \to D$; which will usually be negative and will introduce delays into the regulatory process; and (4) a genetic feedback loop, $A \to G \to T \to D$; which may be positive or negative and will also introduce time delays. In some population systems the last two loops may be inoperative, or may only operate at certain times or at particular population densities. For example, in the case of the Atnarko River salmon run (Figure 3.6), the environmental (or genetic?) feedback loop was apparently operating when the population was at its high-density equilibrium and this created time delays, which resulted in the cyclic trajectory we observed. However, after the population's drastic collapse to its low-density equilibrium, the environmental feedback loop appeared to disengage and the cyclic behavior became less evident. This explanation seems reasonable because we would expect large populations to have more impact on the favorability of their environments than small ones.

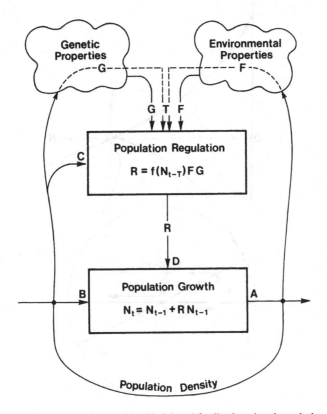

Fig. 3.9 A generalized population model with delayed feedback acting through the environment or gene pool shown as a broken line

3.4. Analysis of the Model

The model defined in Figure 3.9 can exhibit an astounding array of dynamic behavior depending on the form of the density-dependent function and on the presence or absence of environmental or genetic feedback. Because of these complexities, a rigorous mathematical analysis such as we performed in Chapter 2 is impossible. The global stability properties of the model are particularly difficult to evaluate because they depend on the exact form of the density-dependent relationship. However, we can evaluate the local stability of the model in the neighborhood of its equilibrium positions. For example, consider the simple density-dependent relationship shown in Figure 3.10: The equilibrium position, K, is determined by the interception of the function with the abscissa $R = 0$. If we assume that the function is approximately linear very close to this intersection, then we can evaluate its steady-state properties in the immediate vicinity of the equilibrium point. In the magnified view of the equilibrium region (Figure 3.10) we can see that a small negative displacement, $-x$, from equilibrium produces a

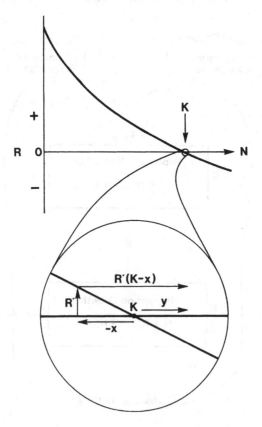

Fig. 3.10 Steady-state analysis of a nonlinear density-dependent function in the neighborhood of its equilibrium point, K, where $-x$ is a very small displacement from equilibrium, R' is the per capita rate of increase at $K - x$, and y is the overshoot of equilibrium following the initial displacement

starting population density of $K - x$ which has a per capita rate of increase of $R' > 0$. Therefore, in the next time increment the population will grow by the addition of $R'(K - x)$ individuals. We can see from Figure 3.10 that the overshoot of the equilibrium point is

$$y = R'(K - x) - x$$

and the overshoot ratio becomes

$$y/x = (R'/x)(K - x) - 1.$$

Now we can also see from this illustration that the slope of the density-dependent function in the neighborhood of equilibrium is defined by $s = R'/x$ which, when substituted in the above equation yields

$$y/x = s(K - x) - 1.$$

Now if we let $-x$ be an extremely small displacement relative to the equilibrium density, K, then we can make the approximation $(K - x) \approx K$, and the overshoot ratio becomes

$$y/x \approx sK - 1. \tag{3.9}$$

The criterion for a stable equilibrium is that the overshoot ratio is less than or equal to unity, and so the system will be unstable when

$$sK > 2. \tag{3.10}$$

As we would expect, this result is identical to that which we derived in our linear analysis (Chapter 2) because, if you remember, $sK = R_m$ in the linear case.

By the same reasoning we can also include the effects of time delays (see Chapter 2), in which case the system becomes unstable when

$$sKT > 2. \tag{3.11}$$

Remember, because we are now dealing with nonlinear equations, an unstable system will not necessarily oscillate to extinction as the linear model predicts. The global stability properties are determined by the overall shape of the density-dependent function remote from the point of equilibrium. In fact, we should suspect that most biological systems have evolved globally stable properties because otherwise they would have gone extinct a long time ago. In the event that the population system is globally stable, but unstable in the neighborhood of its equilibrium point, we are likely to see rather unusual dynamic behaviors, including aperiodic oscillations, cycles, and periodic outbreaks (see references in Note 2.8).

Having examined the stability properties of the equilibrium point created by the dominance of competitive interactions, it is now time to look at cooperative equilibria.

In contrast to the competition curve, which passes downwards through the zero growth line $R = 0$ because s is negative, the cooperation curve passes upwards through this line (Figure 3.11). For this reason equilibria created by cooperative interactions are always locally unstable and any small displacement from the equilibrium will result in continuous growth or decay away from it. For instance, we can see from the magnified view of the equilibrium point (Figure 3.11) that a small negative displacement, $- x$, gives a negative per capita rate of increase, $- R'$, and the population declines in the next increment of time by the quantity $- R'(K_1 - x)$. In other words, a slight displacement below the cooperative equilibrium produces a further population decrease and so on until the population becomes extinct. We can see from the figure that the second displacement from equilibrium is $y = R'(K_1 - x) + x$, and that the ratio is $y/x = s(K_1 - x) + 1$. Allowing that $- x$ is a very small displacement, then the criterion for instability is that $sK_1 > 0$. Thus, as long as the equilibrium position and interaction coefficient are greater than zero the equilibrium

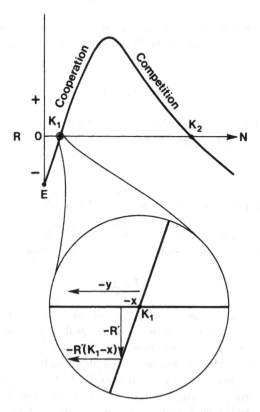

Fig. 3.11 Steady-state analysis of an equilibrium point, K_1, formed by the dominance of cooperative interactions, where $- x$ is a very small displacement from equilibrium, $- R'$ is the per capita rate of increase following this displacement, and $- y$ is the second order displacement after a further time increment

point will be unstable. It therefore becomes apparent that the only stable condition is the extinction of the population.

The unstable cooperative equilibrium (K_1 in Figure 3.11) acts as a dividing line, or a *threshold*, which separates two distinct patterns of dynamic behavior: to the left of this point the population declines to extinction, while to the right it grows toward the upper competitive equilibrium, K_2. Thus, K_1 specifies the *extinction threshold* because it represents the limit to which a population can be reduced before it automatically declines to extinction. This, of course, is a very important concept for a population manager. Perhaps the blue whale population illustrated in Figure 2.4 has already been pushed beyond this threshold?

Another way to view the unstable cooperative equilibrium is that it separates the system in two *domains of attraction* to particular equilibrium points. Considering extinction as a stable equilibrium, E, because once it is reached the system remains there forever, then we specify the domain of attraction to E by N in $(0, K_1)$ and that to the upper equilibrium K_2 by N in (K_1,∞). Once again we see that the domains of attraction are separated by the unstable cooperative equilibrium, K_1.

When we progress to more complicated density-dependent relationships, for instance Figure 3.7C, we find the dynamic behavior of the system is defined by three domains of attraction: (1) extinction behavior for N in the domain $[0, K_1)$, (2) low-density equilibrium behavior for N in the domain (K_1,K_3) and (3) high-density equilibrium behavior for N in the domain (K_3,∞). Again, the behavioral domains are separated by the unstable cooperative thresholds K_1 and K_3.

The concept of domains of attraction to different equilibrium positions is extremely important to those involved in the management of renewable resources. These domains define the boundaries of *resilience* of the system to changes induced by the manager or, for that matter, to any abrupt or gradual environmental changes (see also Note 3.9). In other words, the population can be manipulated, say by harvesting, within a particular domain and it will return to its original equilibrium position when harvesting is discontinued. However, when the harvest is too great, or if harvesting plus an environmental catastrophe forces the population into another domain, then the original equilibrium population may never be attained even if harvesting is discontinued. Thus, the resilience of the system defines the limits to which it can be manipulated and still return to its original condition, and systems that are very resilient will have broad domains of attraction to their equilibria. Once resilience thresholds are exceeded, however, radically different dynamic behavior is initiated, which may be very undesirable from the manager's point of view; for example, the collapse of the Atnarko river salmon run and outbreaks of tree-killing bark beetles.

3.4.1. Environmental and Genetic Effects

Up until now our analysis has been restricted to populations with fixed genetic structure living in constant environments. However, we have argued that these factors will act in concert to determine the amplitude of the basic density-dependent relationship. For instance, the influence of environmental favorability on a simple density-dependent function is shown in three dimensions by Figure 3.8. We can see from this figure that higher equilibrium densities will be possible in more favorable environments. In addition, our stability analysis suggests that population systems will be less stable in more favorable environments because stability in the neighborhood of equilibrium is partly dependent on K, the equilibrium density [see equation (3.10)]. This is a rather important observation because it implies that environmental improvements, say to produce greater crop yields (larger K), may create unstable systems and cause serious management problems. For example, modern agricultural practices enable the farmer to produce very high yields but, at the same time, problems from pest organisms (insects, fungi, nematodes, etc.) arise. Of course, these pests are natural components of the density-dependent regulatory mechanism acting on the crop population and – in the absence of pesticide applications – they would cause highly unstable conditions because outbreaks would periodically decimate the crop populations. Environmental manipulations should, therefore, be carefully studied before they are implemented, with the understanding that a less stable system is likely to be created.

It should be noted at this point that our concept of environmental favorability is an inclusive one. That is, we have included all environmental factors under the general heading of favorability. Although most of the following arguments will be centered around this simplified concept, we should be aware that the environment is composed of a complex set of interacting factors which may affect the organism in different ways. For example, food and space may be directly responsible for setting the equilibrium density for the population, or carrying capacity, while factors such as temperature and moisture may have greater direct effects on the rate of growth towards equilibrium. However, the latter may also affect the equilibrium density indirectly through the rate of food replacement. Predators, pathogens, competitors, and cooperators present in the environment will also affect growth rates and equilibrium densities, and climatic factors may act to moderate these interspecific interactions. Thus, although a change in the favorability of a population's environment may be caused by one or more of these factors, the result will usually be a change in the density of the population at equilibrium and its stability around the equilibrium point.

Of course, genetic changes may also affect the amplitude of the density-dependent relationship in a similar way to the environment. Genotypes with higher reproductive potential or greater survival value, or which allow higher equilibrium densities to be attained, could promote instability in the population system. Thus the population manager should exercise the same caution when he manipulates the genetic composition of his stocks as he does in changing their environments. Although we will be primarily concerned with the dynamics of populations inhabiting changing environments in the remainder of this book, we should always bear in mind that genetic variations may cause similar or even more diverse dynamic scenarios.

3.5. Populations in Changing Environments

When we include environmental favorability as a variable in the density-dependent relationship we obtain a three-dimensional function like that in Figure 3.8. As you can imagine, it becomes rather difficult to analyze the dynamics of a three-dimensional model and so we will reduce the relationship to two dimensions by suppressing the vertical axis. For instance, if we view the three-dimensional function of Figure 3.8 from directly overhead we see that the zero growth plane (the plane of $R = 0$) is divided into two parts by the diagonal equilibrium line, K (Figure 3.12) (see also Note 3.10). To the left of this diagonal the individual rate of increase, R, is greater than zero and the population will grow in this zone, while to the right $R < 0$ and the population will decline if it resides in this zone. Thus, wherever a particular population is situated in this growth space, its qualitative behavior is determined by its position relative to the equilibrium line. As this graph also illustrates the net reproduction, or the change in population density over a given unit of time (i.e., $N_t - N_{t-1} = RN_{t-1}$), we will refer to it as the *reproduction plane*.

We can also include some important information on the qualitative stability properties of the diagonal equilibrium line. Knowing that the neighborhood stability of any point on this line is relative to the equilibrium density at that point, we can divide the line into three sections: a lower section, where K is small and $sK \leq 1$, will exhibit asymptotic stability; a middle section, with $1 < sK < 2$, will be damped-stable;

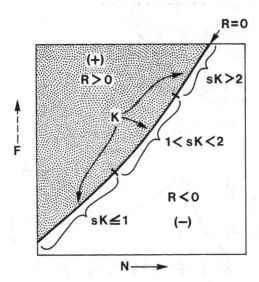

Fig. 3.12 A reproduction plane divided into zones of population growth ($R > 0$) and decline ($R < 0$) by the diagonal equilibrium line ($R = 0$). The density of the population at equilibrium (K) changes in direct relation to the favorability of the environment (F). The equilibrium line is further divided into three sections with different stability properties; in the lower section $sK \leq 1$ providing asymptotic stability, in the midsection we have damped stability because $1 < sK < 2$, and in the upper section the population is unstable because $sK > 2$

and an upper section, with large K and $sK > 2$, will be unstable in the neighborhood of equilibrium (Figure 3.12).

We are now in a position to look at the dynamic behavior of a population on the reproduction plane. For instance, suppose that a population starts at a small initial density, N_0, in a consistently favorable environment (Figure 3.13A). Being to the left of the equilibrium line the population will grow from N_0 to N_1 during the first time interval, then to N_2, and so on. However, because the net growth during any interval of time is dependent on both N and R ($N_t - N_{t-1} = RN_{t-1}$) the magnitude of the growth increments will have to be proportional to these quantities. In the first time increment (Figure 3.13A) population change (RN) was fairly modest because N was small, but in the second period growth was considerably higher because both R and N were relatively large. Of course, as the population approaches the equilibrium line the growth rate must again decrease because R approaches zero. However, the dynamic behavior around equilibrium will be determined by the properties of the line at that level of environmental favorability. In our example the population approaches equilibrium with damped-stable oscillations because it is in the region where $1 < sK < 2$ (Figure 3.13A).

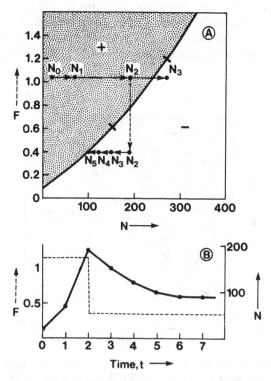

Fig. 3.13 (A) A population trajectory on its reproduction plane showing growth over three time increments (N_0 to N_3) in a consistent environment and also following an environmental deterioration (broken line) at the end of the second time period. (B) A time-series plot of this population trajectory (solid line) with environmental favorability shown as a broken line

Let us now consider the dynamics of this population following a sudden change in the favorability of its environment. Suppose that the environment became less favorable at the end of the second time period (Figure 3.13A, broken line). The population at N_2 is now to the right of the equilibrium line and so it will decline during the next time periods. In addition, because it has been carried into a different stability region, it will now approach equilibrium asymptotically.

We have shown that the reproduction plane forms a useful platform for evaluating the dynamics of populations inhabiting variable environments, and we will use this concept extensively in the remainder of this book. However, we should make it clear that the environment must be assumed to change in discrete steps. In other words we are assuming that the environment remains constant within each time period, but can change at the beginning, or end, of any time increment (Figure 3.13B). Although this assumption may restrict the application of graphical reproduction analysis, it appears quite reasonable for population systems that have discrete life cycles (many insects, salmon, etc.) or that are affected by seasonal patterns that show distinct year-to-year or season-to-season variations.

3.5.1. Environmental Feedback

We have avoided, up until now, the problem of populations affecting the properties of their own environments and the time delays that this may create. However, we should be able to evaluate feedback through the environment by using the reproduction plane. For instance, consider the system depicted in Figure 3.14, where the reproduction plane is divided vertically into two areas by the critical population density, N_c. To the left of this density the population is too sparse to affect the favorability of its environment because resources are renewed as fast as they are used up, but to the right the population uses resources faster than they can be replaced. In effect, the critical density, N_c, represents that population density, which utilizes resources at the same speed as they are produced or regenerated. Now suppose we start with a small population, N_0, growing in a constant favorable environment. We will obtain the horizontal trajectory $N_0 \rightarrow N_4$ (Figure 3.14A). However, as N_3 is above its critical density, the environment for N_4 will be less favorable and so the trajectory will deflect downwards according to the magnitude of the environmental change (broken arrow in Figure 3.14A). The population at N_4, being to the right of its equilibrium line, will decrease to N_5 but, being even denser than N_3, it will reduce the favorability of the environment even further. Continuing with this line of reasoning we will generate the circular trajectory shown in the figure. Notice that as the population density falls below the critical density, N_c, the resources are able to regenerate and the environment becomes more favorable again.

When populations affect the favorability of their own environments they tend to follow cyclic trajectories (Figure 3.14B). These cycles result because time delays are introduced into the regulatory process when population density at the beginning of one time period affects the favorability of the environment in the next. However,

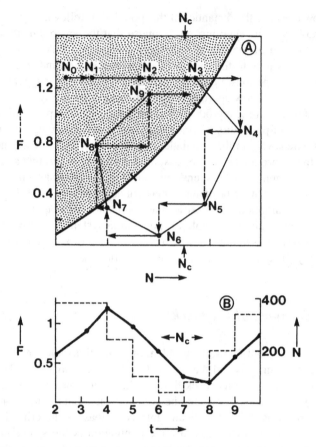

Fig. 3.14 Analysis of a reproduction system in which population densities greater than N_c reduce the favorability of the environment in future time periods: (A) A population trajectory (N_0 to N_9) where the magnitude of population and environmental changes are shown as solid and broken arrows, respectively. (B) A time-series plot of these population and environmental changes

we can see that the time delay is not a simple unit delay but has cumulative or historical aspects. For example, the environment of the population existing at N_6 (Figure 3.14A) is very unfavorable, not only because of its density in the previous time period, but in the preceding two as well; that is, the environmental deterioration was brought about by the combined action of N_3, N_4, and N_5. This is an interesting observation because it implies that the system has cybernetic qualities, or the capacity to store information in the form of memory, and that its behavior depends on "remembered" events, which happened in the past. Although giving population systems the attribute of memory may be stretching a point, there are obvious analogies in the feedback structure of the animal brain.

From a manager's point of view, the dependency of the system's dynamics on historical events has some important implications. Naturally, it is impossible for the manager to change history and, therefore, such systems are difficult to manage without long-term planning and the methodology needed to project the consequences of management decisions into the future. This lesson is extremely important when we consider the impact of expanding human populations on the qualities of their environments. Pollutants released into the environment may pose a greater hazard to future generations than to those, which produced them because they accumulate with time, or their effects are fed back with a delay through complicated ecological pathways (remember the ozone layer, the "greenhouse effect," and the cumulative impacts of the insecticide DDT). You can see some possible consequences of deterioration of environment on population growth, if you run one of the computer examples in the disk that comes with this book (section 2.4.7, Deterioration of environment, and the corresponding model).

Our analysis of environmental feedback on the reproduction plane also helps to explain why certain populations, such as the larch budmoth's (Figure 2.9), cycle in some environments but not in others. We can see that the population in Figure 3.14 would not have cycled if it was living in a much less favorable environment; that is, in an environment where the equilibrium density was lower than the critical density, N_c. Under these conditions the population would have grown to equilibrium asymptotically or, at most, with damped oscillations (the student is encouraged to demonstrate this using Figure 3.14A).

3.6. Complex Density-Dependent Relationships

So far we have only been concerned with reproduction planes created when competitive interactions dominate the population system at all densities (e.g., Figure 3.7A). It is now time to examine systems in which cooperative interactions dominate over particular density ranges. For instance, if the reproduction of individuals declines when population density gets very low because of difficulties they have in locating their mates, we will obtain a unimodal density-dependent relationship similar to that in Figure 3.7B. Now if the amplitude of this curve decreases as the environment becomes less favorable, then it is easy to visualize how the relationship will appear as the environment gradually deteriorates. The "hump" of positive growth (the region above the $R = 0$ line in Figure 3.7B) will slowly decrease until it disappears below the zero growth plane, much like a smooth headland dropping gently into the ocean. From overhead the reproduction plane will look like that in Figure 3.15A. As before, the plane is divided into zones of population growth ($+R$) and decline ($-R$) by a U-shaped equilibrium line. However, the stability properties of this line have been changed considerably by the dominance of cooperative interactions at low population densities. These cooperative processes cause the equilibrium line to swing upwards to the left (shown as a dotted line in Figure 3.15A). As we know, equilibria created by cooperative interactions are inherently unstable (Figure 3.11), and so the left arm of the U-shaped equilibrium line represents an unstable threshold, which separates extinction behavior from the zone of population growth. Whenever the population is below this threshold, or whenever the environment becomes so

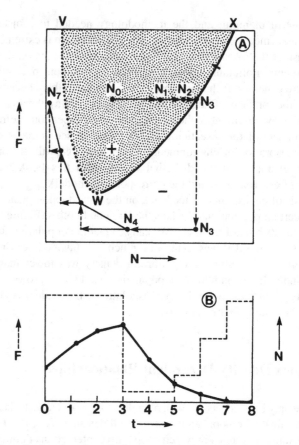

Fig. 3.15 (A) Reproduction plane formed when cooperative interactions dominate at low population densities, showing an extinction trajectory caused by a severe short-term environmental disturbance (the unstable portion of the equilibrium line is shown as the dotted line *V,W*). (B) A time-series plot of the trajectories above

unfavorable that the population is carried below the apex (*W* in Figure 3.15A), then it will decline automatically to extinction. For example, the trajectory in Figure 3.15A shows a population growing asymptotically to equilibrium in a favorable environment, but then being subjected to a severe environment for two time periods. This environmental catastrophe was sufficient to drive the population around the extinction apex and eventual extinction was not prevented by a gradual improvement of the environment (Figure 3.15B). This example illustrates the point that systems dominated by low-density cooperative interactions are extremely sensitive to severe environmental disturbances, such as overharvesting, dam construction, climatic alterations, or other man-made or natural disasters, and that extinction may not be avoided even if efforts are made to rectify the disturbance. It also underscores the logic of hatchery operations where large numbers of organisms are artificially reared to re-stock the declining population. For example, if the population in Figure 3.15 had been re-stocked in the sixth or seventh time periods, so that its density was raised above the unstable threshold, then its extinction could have been prevented. However, if the environment had not improved, then the population could only have been sustained by repeated re-stocking.

In some population systems cooperative interactions may dominate at fairly high densities as well as at low ones, giving rise to complex bimodal density-dependent relationships (Figure 3.7C). In a gradually deteriorating environment the two "humps" of this curve will decrease and eventually disappear below the equilibrium plane to create a W-shaped equilibrium line (Figure 3.16A). This line will be made up of two unstable sections (V,W and X,Y) and two potentially stable sections (W,X and Y,Z). Population systems obeying this kind of reproduction plane may exhibit any one of three basic behavioral patterns: (1) extinction behavior if the density is below the unstable threshold (V,W) or if it is pushed around the apex (W) by unfavorable environmental conditions; (2) low-density equilibrium behavior along the section (W,X) if population density is between the unstable thresholds (V,W and X,Y) and below the apex (X); (3) high-density equilibrium behavior around (Y,Z) if the population is above the unstable threshold (X,Y) and above the apex (Y) or (X).

Population systems characterized by the dominance of cooperative interactions at relatively high population densities exhibit extremely interesting dynamics in slowly changing environments. For example, suppose we have a population in low-density equilibrium, say at point A in Figure 3.16A, and the environment improves very gradually. The equilibrium point will move slowly up the equilibrium line towards the apex (X). However, once it reaches this apex it will enter the domain of the upper equilibrium line (Y,Z) and so it will grow rapidly towards the point B. The change in the environment may be so gradual that it is hardly noticeable, yet it results in a sudden and dramatic alteration in the behavior of the system (see also Note 3.11).

It is also interesting to introduce the concept of environmental feedback into this complex system. When populations are held in the domain of the lower equilibrium line they will, in all probability, be below the critical density (N_c) where environmental feedback is initiated. Hence, time delays are probably minimal and the population should be held in a "tight" equilibrium. However, once populations enter the domain of the upper equilibrium line they are more likely to exceed this critical density, time delays will be introduced into the negative feedback loop, and cyclic trajectories may then be observed (Figure 3.16). In some cases the environment may be so severely affected by the exploding population that it will be carried around the apex (Y) and collapse back down to the lower equilibrium line. In extreme cases the population may even be carried around the apex (W) and become locally extinct.

We have seen that sudden and dramatic changes in the behavior of a population can be initiated by gradual changes in the favorability of the environment. Not so obvious, perhaps, is the fact that similar changes can be triggered by immigration. For instance, a population in equilibrium at A (Figure 3.16A) can be moved across the unstable threshold (X,Y) if large numbers of individuals migrate into the area from surrounding regions and raise the density of the resident population to D.

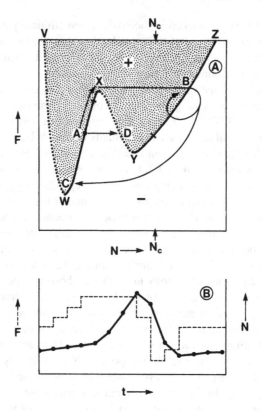

Fig. 3.16 (A) Reproduction plane formed when cooperative interactions dominate at low and intermediate densities, showing several dynamic trajectories in a slowly changing environment (see text for explanation). (B) A time-series plot of the trajectory *A,X,B,C,A* above

Because populations governed by complex reproduction systems can exhibit sudden and unexpected changes in behavior, in response to rather minor environmental disturbances or immigrations, they pose severe problems for the population manager. For example, if the organism described by Figure 3.16 was a pest, say a forest insect, we can see that a destructive outbreak may suddenly erupt for no apparent reason. The environmental change may be so gradual that the manager is unaware of it, or it may even have been caused by the actions of the manager himself. In addition we can see that, once the outbreak has started in a certain area, it can spread rapidly into adjacent regions as immigrants raise local population densities above the outbreak threshold. It is extremely important, therefore, for the manager to control the favorability of the pest's environment and to treat areas of high favorability with extreme caution.

Of course, if the population in question is a useful resource, rather than a pest, the manager would be concerned with keeping it within the domain of the upper equilibrium line. He should be particularly careful not to overharvest the population and thereby push it into the domain of the lower equilibrium line. Even if he is practicing a conservative harvesting strategy, however, the population can still be pushed over the unstable threshold by inclement environmental conditions. This is probably what happened to the Atnarko River salmon run when it was subjected to harvesting and inclement weather during the same year (Figure 3.6). Once again, the manager should be primarily concerned with maintaining the favorability of the environment. In fact, if he can keep environmental favorability above the apex (X in Figure 3.16A), then the population will always be in the domain of the upper equilibria. It is also evident that useful populations can sometimes be re-established at their former levels of abundance through carefully designed stocking and habitat improvement programs. However, the latter is the only really effective way to eliminate the problem of undesirable low-density equilibrium behavior.

Although the theory of multiple equilibrium systems seems well founded and certainly has some useful applications in population management, they are not easy to define for real-life systems. In order to predict the behavior of these systems we need to define the unstable portions of the equilibrium lines. However, because unstable equilibria are transient phenomena, they cannot normally be defined by empirical observations alone. Thus we need to understand the causes of the unstable behavior; that is, how cooperative interactions change with population density. Unfortunately, most of the theoretical and experimental research in the past has concentrated on competitive interactions and the action of mortality factors such as predators, parasites, diseases, and the like. More research needs to be devoted to the ways animals cooperate to avoid these mortality factors and to obtain food. Only then will we gain the knowledge necessary to manage these complex population systems.

We are now ready to leave our analysis of populations consisting of a single species occupying a defined geographic area, and to proceed to more complex systems involving two or more species and large areas in space. However, the concept of the reproduction plane developed in this chapter will be extremely useful in this pursuit. Hence it is important that these ideas are understood by the student. Our analysis has, it is hoped, demonstrated that the general population model we constructed in this chapter performs rather well at simulating the diverse behavior that we observe in nature. Whether this model is accepted by the reader as a valid representation of real life depends, of course, on whether he has been convinced by the arguments of these first three chapters. Whatever, it is now time to proceed to the challenging task of evaluating interacting population systems.

3.7. Chapter Summary

In this chapter we examined the mechanisms involved in the density-dependent regulation of single-species populations, built a general model of the population system, and evaluated the behavior of the model under a variety of conditions. The main points are summarized below:

1. Population growth is regulated by density-dependent feedback acting on the processes of birth, death, and migration. Both negative and positive feedback may be involved in the regulatory mechanism.

   - Negative feedback operates through competition for environmental resources, usually food or living space, which results in higher death rates due to starvation, cannibalism, predation, or disease, lower birth rates caused by malnutrition, disruption of sexual behavior, or lack of nesting places, and higher emigration rates. These effects are transmitted directly and rapidly back to the population, which stimulated them.
   - Delayed negative feedback occurs when populations affect their genetic or environmental properties because the intensity of competition is then dependent on population density in previous time periods. Populations may affect their environments by removing resources faster than they can be replaced, by encouraging the immigration and reproduction of predators, parasites, and diseases, and by polluting their living space. Genetic feedback occurs when certain genotypes, with different reproductive or survival characteristics, are selected for at different population densities.
   - Positive feedback operates when cooperative interactions are important in determining the reproduction and survival of individuals. These interactions, which may involve mating, food capture, escape, or other social behaviors, often dominate at low population densities but, in some systems, they may also become dominant at relatively high densities.

2. The overall effect of feedback regulation was viewed by plotting the individual rate of increase in relation to initial population density. This relationship was rarely linear. Circular plots indicate the presence of time delays in the regulatory mechanism. The length of the delay can often be found by plotting the individual rate of increase against population density in previous time periods. Complicated relationships, with one or more peaks, are found in systems where cooperative and competitive interactions dominate at different population densities.

3. A general population model was constructed to account for the variety in the natural regulatory mechanisms. The basic density-dependent function is moderated by environmental and genetic properties, largely through their effect on the amplitude of the function.

   - The steady-state behavior of the model is dependent on the presence of cooperative and competitive processes, the slope of the density-dependent function, and the population density at equilibrium. Equilibria created by cooperative processes are always unstable, acting as thresholds, which separate distinct patterns of behavior, or domains of attraction to different potentially stable equilibria. Competitive equilibria, on the other hand, may be stable or unstable, depending on the slope of the density-dependent function and the density of the population at equilibrium. The global stability properties, however, depend on the overall form of the density-dependent relationship.
   - Environmental and genetic properties affect the amplitude of the density-dependent function and, through this, the equilibrium density and the stability of

the system. Populations probably evolve a genetic structure that guarantees a globally stable equilibrium under the most prevalent environmental conditions.

4. The dynamics of populations inhabiting variable environments was evaluated on a reproduction plane, which shows the relationships between equilibrium densities, or an equilibrium line, and environmental favorability. Time delays, which are introduced when population density exceeds a critical level where the favorability of the environment is affected, were also evaluated and found to produce population cycles. The environment also plays a crucial role in determining the presence or absence, and the amplitude, of these cycles.

5. Equilibrium lines become quite complicated when cooperative as well as competitive interactions dominate at different population densities. Unstable thresholds may be produced when cooperative interactions dominate, and these separate domains of attraction to potentially stable equilibria, or define the boundaries of population resilience. Dramatic changes in the behavior of the population may occur when these unstable thresholds, or boundaries, are transcended. Low-density cooperative interactions may create extinction thresholds, whilst high-density cooperation can result in systems with two or more potentially stable equilibria. In the latter case, time delays are likely to occur in the regulatory mechanism as populations near their high-density equilibria, resulting in population cycles, collapses to the low-density equilibria, or even local extinctions.

Exercises

3.1. Syunro Utida (see Figure 3.4 for reference) performed an experiment in which he grew populations of pea weevils on 10 grams of Azuki beans (50–60 beans). Starting with 16 weevils he counted their progeny at the end of each generation and then supplied them with a similar quantity of beans. He obtained the following census over nine generations: 16, 294, 125, 250, 130, 213, 160, 200, 150, 180.

  A. Calculate the realized per capita rate of increase for each generation and graph it as a function of population density at the beginning of each generation.
  B. Determine the neighborhood stability of this system by graphical means and by measuring the slope of the curve and the equilibrium density.
  C. Do cooperative interactions play an important role in this system and are time delays present in the density-dependent relationship? If not, why not?

3.2. The density of a population is observed to change over a 10-year period in the following way: 200, 320, 100, 400, 30, 80, 340, 70, 300, 160, 370.

  A. Calculate the individual rate of increase for each year and graph it as a function of the initial population density.
  B. What is the neighborhood stability of this population system? Define the steady-state characteristics graphically and by measuring the slope of the curve and the equilibrium density.

C. What do you think the global stability properties of this system are?

D. Are cooperative interactions and/or time delays operating in this system?

3.3. Over the years 1964 to 1971 a bark beetle population was measured and the following densities were found: 100, 303, 1267, 1333, 832, 212, 157, 321.

A. Calculate the per capita rate of increase for each beetle generation (the beetle has one generation per year) and then plot it against the initial densities; because of the large numbers involved you will obtain a better plot if you transform the data to logarithms.

B. Are time-delays present in this system and, if so, what is the approximate length of the delay?

3.4. Carefully examine Figure 2.6 (page 32).

A. Explain why the equilibrium levels, and the amplitude of the oscillations, are different in oak and pine woods.

B. Do you think time delays are involved in the regulation of these populations?

3.5. Carefully examine Figure 2.7 (page 33) and explain the dynamics observed using a reproduction plane with stand density as your environmental favorability axis: you will have to use your imagination because the equilibrium line cannot be specified exactly by the data.

3.6. Extract the data for the hare and lynx populations from Figure 3.2; that is, calculate the approximate numbers of hares and lynx for each year. Assuming that the hare population size determines the favorability of the lynx's environment, plot the lynx–hare trajectory on a reproduction plane and put in the equilibrium line. You can do this by plotting the net reproduction of the lynx in each year (i.e., $N_t - N_{t-1}$) next to the appropriate hare density at the beginning of the year to give the population change vector in an environment of given favorability. The environmental change vector, then, is the net reproduction of hares.

3.7. Examine Figure 3.6 and explain the observed dynamics of the salmon population by constructing an appropriate reproduction plane.

Notes

3.1. A great deal of semantic confusion has surrounded the concept of density-dependence, which was first introduced by H. S. Smith (*Journal of Economic Entomology*, vol. 28, p. 873, 1935) to describe a mortality factor which destroys an *increasing percentage* of a population as its density increases. In this book we use the term in its broadest sense to mean a feedback mechanism, which responds to the density of the population. This does not necessarily imply negative feedback, as Smith's definition does, and is, therefore, more in line with Haldane's concept (*New Biology*, no. 15, Penguine Books, London, 1953). The various interpretations of the term "density-dependence" are summarized by M. E. Solomon in the book *Natural Regulation of Animal Populations*, edited by I. A. McLaren, Atheron Press, New York, 1971. This book also deals with some of the more recent theories for the natural regulation of animal populations, including genetic feedback and co-evolution. A good summary of the great debate of the fifties, concerning the role of physical versus biological factors, can be found in the book *The Ecology of Insect Populations in Theory and Practice*, by L. R. Clark, P. W. Geier, R. D. Hughes, and R. F. Morris, Methuen & Co., Ltd., London, 1967.

The comprehensive theory of natural control resulted from the work of many ecologists. However, the paper by C. B. Huffaker, which he presented at the *Tenth International Congress of Entomology* in 1958, is one of the first definitive statements leading to the contemporary viewpoint (the paper can be found in the Congress Proceedings, vol. 2, p. 625).

3.2. Examples of unusual behavior brought on through physiological and psychological stress amongst animals living under crowded conditions can be found in Desmond Morris' article "Homosexuality in the ten-spined stickleback" (*Behavior*, vol. 4, p. 233, 1952), and John Calhoun's "Population density and social pathology" (*Scientific American*, February 1962). The latter is based on experiments with crowded rats. Some typical responses to overcrowding were hypersexual and homosexual behavior, cannibalism of young, and continuous fighting amongst dominant males. Robert Ardrey, in his book *The Social Contract* (Athenum Press. New York, 1970), drew parallels between these experiments and the behavior of crowded humans. Although these works have received considerable criticism, their insights should not be dismissed lightly. As Thomas Malthus emphasized so strongly in his *Essay on the Principle of Population* (see Ann Arbor Paperbacks, University of Michigan Press, 1959, for a more recent edition) almost 200 years ago, both misery and vice result from the struggle for scarce resources. Ecologists and demographers have generally been more concerned with misery in the form of starvation, disease, warfare, and such. And rightly so, because modern technology can temporarily alleviate these miseries. However, the experimental work of Morris, Calhoun, and their colleagues, has given weight to the second dimension of the Malthusian thesis, that crowding can cause social stress and lead to abnormal social behavior or, in Malthus' own terms, vice. It is interesting that discussion of vice, and the moral and social issues this gives rise to, is almost as risky today as it was in Malthus' times (see also Note 3.6).

3.3. Dennis Chitty has proposed that antagonistic interactions amongst crowded voles cause changes in the genetic properties of succeeding generations which makes them less resistant to normal mortality factors, and that this delayed genetic feedback is responsible for population cycles. For those interested in Chitty's views, his paper in the *Canadian Journal of Zoology* (vol. 38, p. 99, 1960), and the excellent summary by C. J. Krebs in the book *Population Ecology*, edited by L. Adams (Dickerson Publishing Co., Delmont, California, 1970) are recommended. Chitty also gives a summary of these ideas in the book *Natural Regulation of Animal Populations*, edited by I. A. McLaren (Atherton Press. New York, 1971).

3.4. The view that stability between co-evolving populations of plants, herbivores, and carnivores is maintained by genetic feedback was proposed by David Pimentel. For example, see his contribution to the book *Natural Regulation of Animal Populations* (Note 3.3 for reference). In this sense, the efficient predator (herbivore or carnivore) puts strong selective pressure on its food species to evolve resistance to attack, and this feeds back to the predator population to limit its numbers. Pimentel proposes that, after many such cycles, a stable equilibrium between predator and prey populations is attained; that is, the system approaches equilibrium with damped-stable oscillations. Pimentel also reports on laboratory experiments that support this argument.

3.5. Bark beetles of the family Scolytidae, order Coleoptera, are insects that attack and kill living trees and then reproduce in the dying host. These beetles have evolved a system of chemical communication (pheromones), which draws beetles flying nearby to a recently attacked tree, and this "mass attack" helps them overcome the defenses of their host. When large numbers of beetles are flying, even healthy, vigorous trees can be overwhelmed because the rapid mass attack circumvents the host's defenses that need a period of time in which to operate effectively. When populations are small, however, the tree's defenses are usually effective and the beetle population can only succeed in colonizing unhealthy individuals. Because of this divergent behavior, caused by cooperative activities acting at high population densities, bark beetle systems sometimes have multimodal density-dependent curves like that of Figure 3.6C; one such can be found in the paper by the senior author in the *Bulletin of the Swiss Entomological Society* (vol. 52, p. 227, 1979). For those interested in pursuing this subject we would suggest another of the papers by the senior author (*Bioscience*, vol. 22, p. 598, 1972) which deals with conifer defense systems, and one by J. H. Borden in the book *Pheromones*, edited by M. C. Birch (North Holland Publishing Co., Amsterdam, 1974) which reviews the pheromones of bark beetles and the behavior that they elicit.

3.6. The subject of sociobiology, or the biological basis of social behavior, is one of the newest and most controversial areas in the biological sciences. Its leading proponent, E. O. Wilson, who received the Pulitzer Prize for his book *On Human Nature* (Harvard University Press, Cambridge, 1978) presents this thesis in detail in *Sociobiology: The New Synthesis* (Belknap Press, Cambridge, Mass., 1975). Sociobiologists attempt to explain the origin of social behavior within a framework of classical evolutionary theory. That is, they are interested in the genetic basis of cooperative social interactions (e.g., sex selection, parenthood, and altruism), as well as competitive interactions (e.g., aggression and territoriality). The major controversy arises when these ideas, which arose largely from the study of "lower" animals, are applied to human behavior. The proposition that human behavior is rooted in evolutionary history is repugnant to some scientists who stress that the cultural environment is the dominant force molding human social behavior. Like most scientific controversies, the

truth undoubtedly lies somewhere in between. For those interested in a philosophical analysis of the sociobiological debate, albeit slanted towards Wilson's
views, we would suggest M. Ruse's book *Sociobiology: Sense or Nonsense* (D.
Reidel Publishing Co., Dordrecht, Holland, 1979). Popularized versions of
early sociobiological ideas can also be found in Robert Ardrey's *The Social
Contract* (see Note 3.2) and Desmond Morris' *The Naked Ape* (Dell Publishing
Co, New York, 1967).

3.7. Stephen D. Fretwell, in his book *Populations in a Seasonal Environment*
(Princeton University Press, New Jersey, 1972) develops what he calls a theory
of habitat suitability which is fundamentally similar to our ideas of environmental favorability and density-dependent feedback. Fretwell considers the
suitability of an organism's habitat, or living environment, to be dependent on
the density of the population and the basic properties of the habitat. He explains
that the basic suitability of the habitat defines the maximum individual rate of
reproduction and survival in that habitat when population density is very low,
and that this is reduced in proportion to population density. His idea of basic
habitat suitability is, therefore, equivalent to our environmental favorability.

These views are, of course, a considerable simplification of the qualities of real
environments, which are composed of a complex of physical and biotic properties.
However, environments can often be classified according to their favorability,
or suitability, for particular organisms and, hence, the concept seems to be of
practical use as well as theoretically reasonable. An example is the classification
of forest habitats (environments) according to the types of plants, which are
favored (see Chapter 5 of this book).

3.8. Although we have chosen not to specify the mathematical form of the density-
dependent relationship, a number of nonlinear models have appeared in the
ecological literature. Perhaps the most popular is the exponential decay equation

$$R_0 = \lambda e^{-aN_{t-T}},$$

where R_0 is the per capita replacement rate ($R_0 = N_t/N_{t-1} = R + 1$), λ is the maximum per capita replacement rate ($\lambda = R_m + 1$), which is frequently called the
finite rate of increase, and α is the density-dependent coefficient. This equation
has been used most extensively by fisheries biologists and is often referred to
as the Ricker curve after the well-known fisheries ecologist W. E. Ricker (e.g.,
see his monumental monograph "Stock and Recruitment" in the *Journal of the
Fisheries Research Board of Canada*, vol. 11, p. 559, 1954). The equation can
also be written in terms of carrying capacity

$$R_0 = \lambda^{1-N_{t-T}/K}$$

and can also be generalized to provide a wider range of behavior, following the
ideas of M. E. Gilpin and F. J. Ayala (*Proceedings of the National Academy of
Science USA*, vol. 70, p. 3590, 1973), by the addition of another parameter

$$R_0 = \lambda^{1-(N_{t-T}/K)^{\theta}}$$

Some other models, which are commonly used are due to J. B. S. Haldane (*New Biology*, vol. 15, p. 9, 1953),

$$R_0 = \lambda N_{t-1}^b$$

R. J. H. Beverton and S. J. Holt ("On the dynamics of exploited fish populations," *Fishery investigations*, series 2, vol. 19, printed by Her Majesty's Stationary Office, London, 1957),

$$R_0 = (\alpha + \beta N_{t-1})^{-1}$$

and M. P. Hassell (*Journal of Animal Ecology*, vol. 44, p. 283, 1975),

$$R_0 = \lambda(1 + \alpha N_{t-1})^{-\beta}$$

Those wishing to investigate the dynamic properties of these various models should consult the paper by R. M. May and G. F. Oster (*American Naturalist*, vol. 110, p. 573, 1976).

3.9. Ecological resilience addresses the capability of ecological systems to absorb shocks from external forces and, in particular, their ability to recover from man-made disturbances such as harvesting, pollution, dams, insecticide applications, etc. A resilient system will have a broad domain of attraction to a stable equilibrium so that it tends to return to its original condition following severe disturbances. Such systems are said to be robust. On the contrary, fragile systems have little resilience because of their constricted domains of attraction and even minor disturbances may precipitate movements towards configurations quite different from their original condition. For those interested, the concept of resilience is discussed by C. S. Holling in his article in the *Annual Review of Ecology and Systematics* (vol. 4, p. 1, 1973).

3.10. The equilibrium line can be precisely defined, of course, if the density-dependent function $f(N_{t-T})$ is known, and given a relationship between environmental favorability F and the maximum individual rate of increase R_m. For instance, if we have our familiar linear function, $R = R_m - sN$, then from Chapter 2 we know that

$$K = R_m / s.$$

Assuming that R_m changes linearly with the favorability of the environment, and that $R_m \to 0$ as $F \to 0$, then we have

$$R_m = bF,$$

where b is the benefit of a unit increase in the favorability of the environment. Thus we can write

$$K = \frac{b}{s}F$$

and we see that the equilibrium line is determined by the combined action of the environment and the self-inhibiting density-dependent interactions.

In the case of nonlinear density-dependent functions, such as the exponential relationship

$$R + 1 = R_m e^{-sN},$$

we will obtain nonlinear equilibrium lines. For example, given the exponential relationship above, and transforming to logarithms, we obtain

$$\ln (R + 1) = \ln R_m - sN,$$

where ln refers to the natural logarithm. At equilibrium we get

$$0 = \ln R_m - sK,$$

or

$$K = \ln R_m / s.$$

Assuming that the environment acts linearly on R_m, then we can substitute bF for R_m to yield

$$K = \ln (bF) / s,$$

which will give us a relationship similar to text Figure 3.12.

3.11. These complex W-shaped equilibrium lines bear superficial resemblance and predict similar behavior, to the equilibrium manifolds of "catastrophe" theory, a branch of topology first introduced by the French mathematician Rene Thom in his book *Structural Stability and Morphogenesis* (Benjamin-Addison Wesley, New York, 1975). This theory deals with systems that exhibit abrupt, discontinuous or divergent behavior and has been applied to a number of real-life systems, particularly by Christopher Zeeman (see, for example, his paper in *Scientific American*, vol. 234, p. 65, 1976). However, there seems to be considerable debate amongst mathematicians over the validity of catastrophe theory as applied to the natural sciences (e.g., see the papers by H. J. Sussmann and R. S. Zahler in *Behavioral Science*, vol. 23, p. 383, 1978, and by G. B. Kolata in *Science*, vol. 196, p. 287, 1977). Although the equilibrium systems developed in this book are based on biological arguments, and do not require the notions of catastrophe theory, catastrophe theorists, and particularly Christopher Zeeman have contributed much to our thinking.

Part II
Systems of Interacting Populations

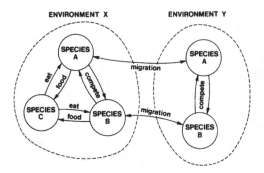

In the first part of this book we considered populations of a single species living within specific geographic boundaries. All other species, and other populations of the same species, were relegated to the environment of the subject population. Although this may be a reasonable approach when we are interested in a particular species living in a certain place, it is often necessary, for practical or academic reasons, to consider the interactions between populations of different species occupying the same space, or between populations of the same species living in different places. For example, forest stands are frequently composed of several intermixed tree species and forest ecologists and managers are interested in the dynamics of these interacting populations. Likewise, the interaction between predator and prey populations is of interest to the ecological theorist and the practitioner of biological control of pests. The spatial interactions between populations of the same species are of particular concern when migrations lead to significant changes in population behavior - for instance, in the spread of a pest insect or of disease epidemics.

In this part of the book we will examine interacting population systems using the basic models and analytical methods developed in Part I. We will first look at interactions between populations of two species living in the same place (Chapter 4), then interactions between populations of the same species living in different places (Chapter 5) and, lastly, communities composed of many interacting populations (Chapter 6).

Chapter 4
Interactions Between Two Species

4.1. Population Interactions

Populations of two different species that coexist within the same geographic area may be viewed as two separate population systems, which interact with each other through their common environment. In this way, the numbers of one population modify the favorability of the environment for the other (Figure 4.1). This interaction creates an additional feedback loop, shown as a bold line in the figure, which passes through both population systems. This loop may be positive or negative depending on the signs of the interspecific interactions.

Different species may interact with each other in a number of ways, depending on whether their presence improves (+), detracts from (−), or has no effect on (0), the environment of the other species. The combined interactions can be specified in an interaction matrix, which describes all possible interactions between two species:

| | | Effect of species A on B's environment | | |
		+	0	−
Effect of species B on A's environment	+	++	0+	−+
	0	+0	00	−0
	−	+−	0−	−−

From this matrix we see six types of interactions that may occur between any pair of species. These are usually termed mutualism or symbiosis (+ +), commensalism (+0), predation (+ −), competition (− −), amensalism (−0), and indifference (00). However, for our purposes these may be reduced to three basic kinds of interactions (see also Note 4.1):

1. One or both species supply a commodity or resource, which is useful to the other but is of no use to its own population. Thus, although the donor species improves the environment for the recipient it does so at no cost to itself. We will call such interactions *cooperative* and they will include mutualism (+ +) and commensal-

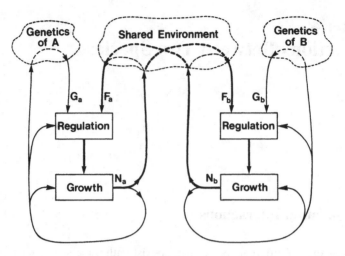

Fig. 4.1 A general model for two interacting species, A and B, where the numbers of the first species, N_a, affect the favorability of the other species' environment, F_b, and *vice versa;* the G's indicate genetic properties of the two species

ism (+0). Examples are ants, which tend aphid colonies for the honeydew they secrete, and in return protect them from predators; and dung beetles, which feed on cattle droppings and thereby increase the area available for grass to grow.

2. One species supplies a resource for the other but, in so doing, its own population suffers. Such interactions involve the feeding of herbivores, predators, parasites, and diseases on populations of their prey. These are grouped under the broad title of *predator-prey* interactions (+ −).

3. One or both species utilize a commodity or resource, which is needed by the other. The result of this is usually competition (− −) between the species for the common resource. A rather rare phenomenon that falls within this category is amensalism (−0), which exists when both species require the same resource but one is excluded from competing for it by some act of the other. Examples of amensalism can be found in certain plants that secrete toxic substances into the soil, preventing other species from invading the site. However, we will consider these under the title of *competitive* interactions.

4.2. Cooperative Interactions

If we examine the interaction feedback loop of Figure 4.1, we will find that cooperative interactions between species will create an overall positive feedback effect; an increase in the density of species A improves the environment of B, and this causes an increase in B's individual rate of increase and population size, which then improves the environment of A and, eventually, its population size, and so on. This feedback loop can be written $N_a \xrightarrow{+} F_b \xrightarrow{+} N_b \xrightarrow{+} F_a \xrightarrow{+} N_a$, and we see that the product of all these positive interactions produces an overall positive feedback loop.

Because the density of one species can be interpreted in terms of environmental favorability for the other, we should be able to evaluate cooperative interactions using reproduction planes. Assuming that all other environmental conditions are constant, then we would expect the favorability of the environment for one species to be proportional to the population density of the other, and the equilibrium lines should appear as shown in Figure 4.2A,B. These lines are drawn so that they intercept the axis of each species at a positive density, which means that both species can exist in the absence of the other; that is, they are not completely dependent on each other, as is the case in the ant-aphid example. The intercept K, therefore, represents the saturation density of each species in the absence of the other, while the

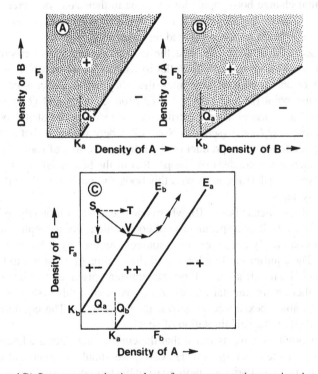

Fig. 4.2 (A and B) Separate reproduction planes for two cooperating species, A and B, respectively, where F is the environmental favorability axis as affected by the density of the other species, K is the saturation density in the absence of the other species, Q is the marginal benefit of the cooperator, and (+) and (−) indicate the zones of population growth and decline. (C) The combined reproduction plane produced by superimposing B's plane on top of A's (see Note 4.3), where both species grow in the (++) zone, A grows and B declines in the (+−) zone, and B grows and A declines in the (−+) zone. The equilibrium line for each species is indicated by E. The particular trajectory ($S,V,\ldots$) is the result of consecutive changes in the density of each species from S to U and S to T, etc., over each time increment

slope of the line Q reflects the improvement to the environment produced by adding a single cooperator. In effect, the slope of the line is the *marginal benefit* provided by the cooperating species to the reproduction and survival of the other species (in Note 4.2 this slope is shown to be defined by p/s, where p is the benefit derived from each cooperator and s is the intraspecific inhibitory effect).

The reproduction plane of species B can now be superimposed on that of A by rotating it clockwise through 90° and then inverting it (see Note 4.3). When we do this we obtain a combined reproduction plane, which is divided into three regions where (1) both populations grow (+ +), (2) species A grows but B declines (+ −), and (3) species B grows but A declines (− +). The dynamics of this interacting system can be evaluated by starting at any position on the reproduction plane and calculating the direction that the system will move over a number of time increments. For example, from point S in Figure 4.2C population A will grow, say to T, while B will decline, say to U, and the net result will be a movement from S to V. After another similar change both populations will be in their zone of increase and they will continue to grow *ad infinitum*. Similar dynamic trajectories can be calculated from any starting point on the combined reproduction plane.

We can see from Figure 4.2C that the interaction produces perpetual growth because the equilibrium lines run parallel to each other. However, if we decrease the slope, Q, of one or both of these lines, then they will intersect to produce a stable equilibrium point (Figure 4.3A). As the product of Q_a and Q_b must be unity when the lines are parallel, then the criterion for a stable coexistence between the two cooperators is $Q_a Q_b < 1$ (see also Note 4.2, where we show that a stable equilibrium will occur when the product of the inhibitory effect of each species on its own rate of increase is greater than the product of the beneficial effect of cooperation). See also the disk that comes with this book for examples of the dynamics of cooperative systems.

The case of commensalism (+ 0), where one species is completely indifferent to the other, reveals itself as a special case of cooperation. For example, suppose that species A is completely dependent on resources supplied by B but that B is indifferent to A. The equilibrium line for A will rise in direct proportion to the population density of B while B's line will remain constant, provided that its numbers are limited by other environmental factors. As Q_b is zero in this system it fulfills the criterion for stable coexistence as shown in Figure 4.3B. The equilibrium is, of course, enforced by the self-limitation of species B.

Up to this point we have assumed that the equilibrium lines are linearly related to the density of the other species. However, we should suspect that this would rarely be true in nature. Perhaps a more reasonable reproduction plane is that shown in Figure 4.4A. In this case the reproduction and survival of one species increases with the density of the other until competition for a diminishing environmental resource brings growth to a halt at a new saturation density, K'. For example, the population may be limited at its lower saturation density, K, by food shortage, and the addition of cooperators to its environment increases this food supply. However, as the density of the population rises, another resource (e.g., nesting places) may run out and limit the population at K'. When reproduction planes of this kind are

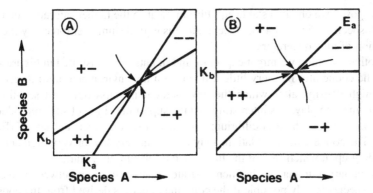

Fig. 4.3 (A) Superimposed reproduction plane similar to Figure 4.2C except that the slope of B's equilibrium line has been decreased, and (B) superimposed reproduction plane for a commensal A, which is completely dependent on B, while B is indifferent to A but is limited by other environmental factors

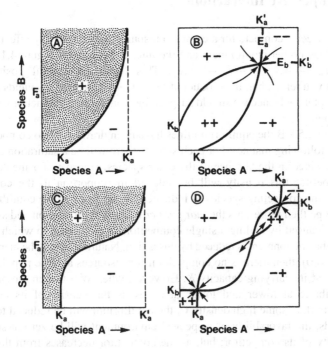

Fig. 4.4 (A) Reproduction plane for a species whose environment is improved by a cooperator until its density becomes limited by other factors at K'. and (B) the superimposed reproduction plane for two such species. (C) Reproduction plane for a species whose environment is only improved after its cooperator reaches a fairly high density and is also limited by other environmental factors, and (D) the superimposed plane for two such species

superimposed we obtain a stable equilibrium near to the higher saturation densities of both species (Figure 4.4B). Once again this equilibrium is enforced by the self-limitation of the cooperators.

Another variation we might expect to find in nature is illustrated in Figure 4.4C. Here, the cooperator has very little influence on the environment until it reaches a fairly high density, after which the environment is improved until self-limiting forces come into play. It is interesting that the superimposition of such reproduction planes may create three equilibrium positions, two of which are stable (Figure 4.4D). The cooperating populations may coexist at very low or very high densities depending upon which side of the unstable equilibrium they start.

In summary, our analysis has demonstrated that equilibrium between two cooperating species is only possible if the combined benefits derived from the cooperative interaction are less than the combined inhibitory effects of each species on their own rates of increase. In other words, the cooperative interaction is basically unstable and stable coexistence is only made possible by the internal negative feedback mechanisms of one or both populations.

4.3. Competitive Interactions

When two species compete for a common resource they negatively affect the favorability of each other environment and the interaction loop of Figure 4.1 becomes $N_a \xrightarrow{-} F_b \xrightarrow{+} N_b \xrightarrow{-} F_a \xrightarrow{+} N_a$. This forms an overall positive loop because, if you remember, the product of two negative interactions yields a positive effect (Chapter 1). Hence we should suspect that competitive interactions are generally unstable.

In Figure 4.5A,B the equilibrium lines have been drawn for two competing species in the following manner: For each species there will be a saturation density K, which is attained in the absence of the other species. However, for the addition of each competitor this density will be reduced in proportion to the competitive strength of the competing species as indicated by the slope of the equilibrium line W. This slope then represents the *marginal cost* to the reproduction and survival of the species caused by adding a single competitor, or the degree to which the favorability of the environment is reduced by this action. Naturally, when the marginal cost is zero ($W = 0$), then there is no competition for resources and the population will equilibrate at its carrying capacity K. However, when $W > 0$ then the population will equilibrate at lower and lower densities as the density of its competitor increases until, at some high density C, the equilibrium will be reduced to zero. In other words, the favorability of a species' environment will be zero at some very high density of its competitor; but, as the competitor decreases from this critical density C, higher and higher equilibrium densities will be possible until, when the competitor is absent, the species will reach its saturation density K.

The equilibrium lines for the two species can be superimposed by inverting one and then rotating the other clockwise through 90° (Figure 4.5C) (see also Note 4.3).

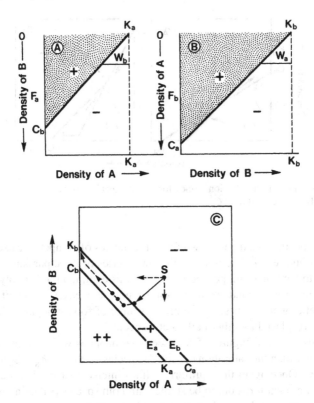

Fig. 4.5 (A and B) Reproduction planes for two competing species, A and B, respectively, where the density of one detracts from the favorability of the other's environment; K represents the saturation density of each species, W the marginal cost of the competitor, C the critical density of one species which reduces the favorability of the other's environment to zero, and (+) and (−) the respective zones of population growth and decline. (C) The superimposed reproduction plane (see Note 4.3 for how it was obtained) with equilibrium lines, E, and a particular dynamic trajectory

In this particular example we can show, by plotting a trajectory or two, that the system will equilibrate at K_b, by which time species A will have become extinct. We notice, as expected, that the interaction between the two species is unstable.

Now suppose we increase the saturation density of species A in Figure 4.5 so that $K_a > C_a$; in doing this we have also increased the marginal cost of species B because W_b is now larger. We now find that the two equilibrium lines intersect to create an equilibrium point (Figure 4.6A). However, this equilibrium also turns out to be unstable because, when the system is displaced from it, it moves towards K_a or K_b depending on the direction of the initial displacement (the student is encouraged to prove this by plotting trajectories on Figure 4.6A).

In both of our examples so far, one species will always dominate the other and eventually drive it to extinction. This is usually termed *competitive exclusion* and it seems to be a fairly common phenomenon in nature; for example, the succession of species that dominate, in their turn, dynamic plant communities (this will be

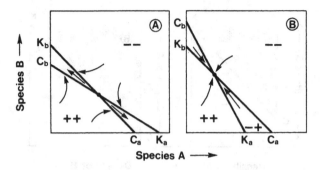

Fig. 4.6 Superimposed reproduction planes for two competing species when (A) $K_a > C_a$ and $K_b > C_b$, and (B) $K_a < C_a$ and $K_b < C_b$

discussed below in more detail). Two further instances of competitive exclusion are illustrated in Figure 4.7. The first shows how a series of three parasites introduced into Hawaii to control a fruit pest competed for this common food supply until only one remained. The second example shows how a weaker competitor, the fir engraver beetle, which is first to colonize a source of food in a particular locality, is eventually displaced by other bark beetle species.

Now let us return to Figure 4.5 and see what happens when we increase C_b so that it is greater than the saturation density of species B; that is, $K_b < C_b$ and $K_a < C_a$ (Figure 4.6B). Once again the equilibrium lines intersect each other, but this time the equilibrium point turns out to be stable with both species persisting in the environment (the student should demonstrate this by plotting trajectories on Figure 4.6B). Thus, *competitive coexistence* is only possible under the conditions that $K_a < C_a$ and $K_b < C_b$. Now as the saturation density K is dependent on the self-limiting effect of the species (i.e., the intraspecific competitive component) while C depends on the strength of the interspecific struggle, then this result means that coexistence is only possible when the repressive effect of each species on its own cohorts is greater than that on its competitor (this result is derived more formally in Note 4.4). This reinforces our conclusion that competitive interactions are inherently unstable and that stability can only be enforced by the self-regulatory mechanisms of the competing species. See also the disk that comes with this book for examples of the dynamics of competitive systems.

Although long-term data demonstrating the competitive coexistence of species living under natural conditions are difficult to find, a number of elegant laboratory experiments have been performed (e.g., Figure 4.8). In this remarkable experiment we clearly see that the populations attain the same equilibrium levels regardless of their initial starting conditions. It also illustrates one of the fundamental principles of competition – the *advantage of numbers*. In Figure 4.8A the more successful competitor (the one that equilibrated at the higher density) started out at a higher density and both populations grew rapidly until they reached equilibrium in about 20 weeks. In the second experiment, however, the weaker competitor was given the advantage of numbers and it was able to maintain a high population density for

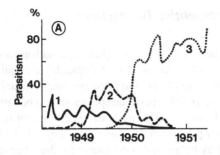

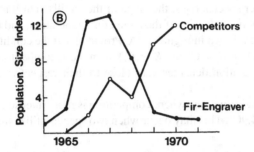

Fig. 4.7 (A) Competitive exclusion of two parasites, introduced into Hawaii to control a fruit fly, by a third species which was introduced last (redrawn from H. A. Bess, R. van den Bosch, and F. H. Haramoto, *Proceedings of the Hawaii Entomological Society*, vol. 17, p. 367, 1961). (B) Exclusion of the fir engraver beetle from fir trees by other bark beetle species that compete with it for food (redrawn from A. A. Berryman, *Canadian Entomologist*, vol. 105, p. 1465, 1973)

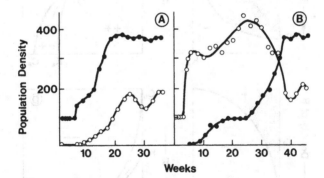

Fig. 4.8 Competitive coexistence between two species of grain beetles living in a fixed quantity of wheat [redrawn from A. C. Crombie, *Proceedings of the Royal Society (Section B)*, vol. 133, p. 76, 1946]

some time before its competitor was able to exert its dominance. Because of this, the equilibrium state was not established for more than 40 weeks. These experiments demonstrate that a weak competitor can temporarily hold an area, in the face of dominant competitors, by occupying the area before the other species arrive.

4.3.1. Nonlinear Competitive Interactions

We should suspect that linear competitive relationships, such as those in Figures 4.5 and 4.6, would rarely be seen in nature. For example, large ungulates like elk and deer utilize a variety of browse plants, and different species usually have different food preferences. Thus, competition between species should be rather inconsequential as long as their populations remain small, because different food plants will be eaten. However, when their populations become large they may be forced to eat less preferred plants, which may be a major food source for the other species, and competition will intensify. Because the effects of competition become stronger as the density of the other species rises, the slope of the equilibrium line will increase in direct relationship to the density of the competing population, and the reproduction plane will appear as shown in Figure 4.9A. Provided that the conditions for coexistence are met (i.e., that $K_a < C_a$ and $K_b < C_b$), then these competitors will come into equilibrium at population densities very close to their respective saturation levels (Figure 4.9B).

We may also find situations where competition is more intense at low population densities (Figure 4.9C). This may occur when two species utilize the same resource,

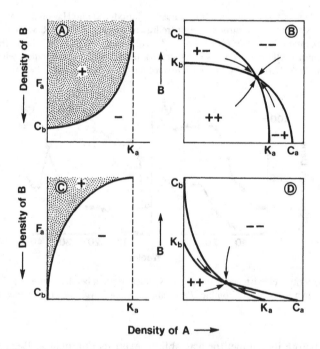

Fig. 4.9 (A) Reproduction plane for a species that is more severely affected by high densities of its competitor; (B) the superimposed reproduction plane for two such species. (C) Reproduction plane for a species that is more severely affected by low densities of its competitor (see Note 4.3 for its construction); (D) the superimposed reproduction plane for two such species

but resort to different utilization patterns if forced to by intense competition. Equilibrium lines with this general form have been observed in experimental fruit fly populations competing for a common food supply in culture flasks (Note 4.5). In this particular experiment the slope of the equilibrium line decreased with increasing density of the competing species; perhaps living with a few flies makes life difficult, but living with a few more when a multitude is already present makes little difference. When species with this kind of reproduction plane interact they will come into equilibrium at rather low densities, provided, of course, that the conditions for coexistence are met (Figure 4.9D).

The shape of the equilibrium line is affected not only by the competitive interaction between the species, but also by the intrinsic density-dependent processes acting on each population. This is demonstrated in Note 4.4, where it is shown that the marginal cost of competition (the slope of the equilibrium line) is $W = c/s$, where c is the effect of the competitor, and s the intraspecific effect. Hence, the shape of the equilibrium lines in Figure 4.9A,C could also be affected if intraspecific competition changes with population density (see also Note 4.5). Under these conditions we can imagine the interaction between species that have differently shaped reproduction planes. For example, the interaction between the planes of Figures 4.9A and C is shown in Figure 4.10. Given the proviso for coexistence, such interactions will equilibrate with one species at a high population density and the other at a low density (Figure 4.10A). In addition, we may also find systems with more than one equilibrium state. For instance, if we increase the carrying capacity of species A in Figure 4.10A so that $K_a > C_a$, then we will obtain the system shown in Figure 4.10B. Here we find a stable equilibrium with both species present but B being much more abundant, an unstable equilibrium close to C_a, and a stable equilibrium with species A alone (i.e., K_a). This system is particularly interesting because species A can be attracted to two equilibrium levels, depending upon which side of the unstable threshold the trajectory begins. This result has implications, which should not escape the population manager. For instance, suppose the system is at K_a with species A only present, and we begin harvesting this species to levels below C_a.

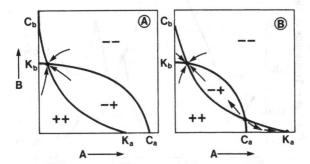

Fig. 4.10 Interactions between competing species with different forms of reproduction plane: (A) Coexistence when $K_a < C_a$ and $K_b < C_b$; (B) two stable equilibria produced when $K_a > C_a$ and $K_b < C_b$

Although species B will be absent at first, once a few individuals obtain a toehold their population will grow until the system equilibrates with B the dominant component, even if the cropping of species A is discontinued. Conversely, if we now switch to harvesting species B, then A will be able to grow slowly and, under prolonged and heavy exploitation of B, it may cross the unstable threshold near C_a, and eventually dominate the system with species B becoming extinct.

These results may help us to explain why certain heavily exploited species do not seem to recover to their normal levels when harvesting is discontinued – a problem which seems particularly acute in the history of herring, sardine, pilchard, and anchovy fishing (see Note 4.6). The relationship between overexploited and declining sardine stocks in California, South Africa, and Japan, and the subsequent rise of the anchovy fisheries is particularly intriguing (e.g., Figure 4.11). However, these observations on natural populations do not necessarily imply that double equilibria, as shown in Figure 4.10B, are present in the system. Even if the system has but a single stable equilibrium, such as that in Figure 4.10A, the overexploited stock may take a long time to re-establish dominance in the face of large numbers of its competitor. The advantage of numbers is, as we have seen, of considerable significance in systems of competing species. In addition, we will find later that there are other explanations for systems with two or more equilibria.

4.3.2. Competition in Variable Environments

Our analysis of systems of competing populations has, of necessity, been conducted under the explicit assumption that all other environmental conditions remain

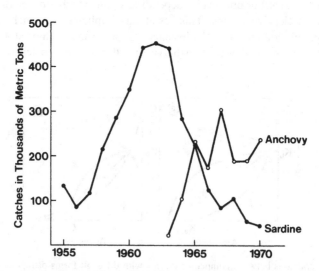

Fig. 4.11 Growth and collapse of the South African sardine fishery and the following rise of the anchovy catch (drawn from data by G. I. Murphy; see Note 4.6 for reference)

constant – that is, all factors other than the densities of the competing species. In this section we will briefly look at the behavior of the system when this assumption is relaxed. We will refer to the environment, excluding the two species, as the *shared environment*, which will include the disputed and undisputed resources, weather, predators, diseases, etc.

We would expect different shared environments to affect the reproduction and survival of two species in different ways and this will, of course, affect the saturation density of one species relative to the other. For example, if Figure 4.9B represents a stable interaction between two species occupying a particular shared environment, then a change in that environment to favor species A may raise its saturation density until an unstable condition is attained when $K_a > C_a$. This is illustrated in Figure 4.12, where we see that the two populations coexist until the saturation level of A exceeds the critical density, C_a, after which B is excluded from the environment.

Now our model (Figure 4.1) also includes a feedback loop representing the effect of a population on the properties of its own environment, an effect that registers in the rate of increase of its own membership. Dense populations will generally reduce the favorability of the shared environment for their own species, but the effect on the competing species may be positive, zero, or negative. Under the condition that environmental changes induced by one species are much more detrimental to its own species than to its competitors, we can explain the phenomenon of *ecological succession*.

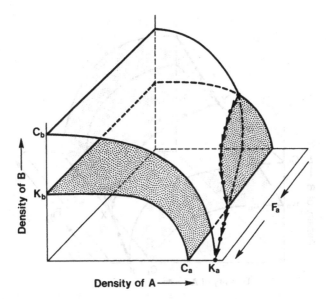

Fig. 4.12 Effect of changes in the shared environment on the interaction between two competing species, where F_a is the favorability of the environment to species A, and F_b is assumed to remain constant. The trajectory shows the path taken by the equilibrium point as the environment slowly changes to favor species A

Suppose we have a shared environment, which is very favorable for species A so that another species B is excluded (i.e., we are in the foreground of Figure 4.13). However, if A causes fairly permanent changes in the environment which are unfavorable to its own members but which favor species B, then the saturation level of A will decline while that of B will increase. If both saturation levels cross their respective critical densities, we will obtain the system shown in Figure 4.13. We can see that, as the shared environment gradually changes in favor of B, the equilibrium density of A decreases until, after the point X, stable equilibria are possible with species B present in the system. At first B will only be present in very small numbers but, as the environment continues to change in its favor, its status relative to A will continue to improve. We can see that both species continue to coexist as long as the saturation level of B remains below its critical density. After this point, Y, species A will eventually be excluded, and the succession of A by B will be complete.

Dynamic scenarios, such as that described above, are very common in the succession of plant communities, and probably just as common, although not so commonly observed, in animal communities. Pioneer species, for example lodgepole pine, colonize environments denuded by fire and other natural disasters. As these denuded environments are favorable for their reproduction and survival, the pine stands may become very dense. However, once the pioneer stand is established, the environment changes drastically. In particular, so little light penetrates the dense

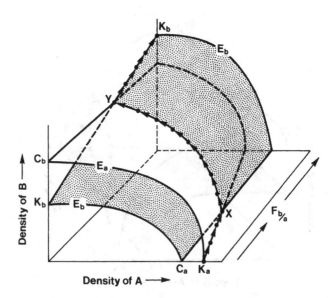

Fig. 4.13 The dynamics of succession: The relative favorability of the shared environment for species B over A, $F_{b/a}$, changes slowly over time causing the equilibrium of the two species interaction to move along the trajectory K_a, X, Y, K_b, where species A outcompetes B on the section K_a, X, both species coexist on the section X, Y, and species B replaces A on the section Y, K_b

canopy that the survival of seedlings is reduced almost to zero – pioneer species, being adapted to colonizing open ground, are not usually very tolerant of shade. However, other more shade-tolerant species, such as firs and spruces, will slowly establish themselves in the understory. At first they will only be present in small numbers but, as the parent pines begin to die and as the successors reach reproductive maturity, their numbers will increase rapidly. Eventually the shade-intolerant pioneers, being unable to reproduce under the dense canopy, will disappear and the succession will be complete.

4.3.3. Strategies of the Competitor

Competition, whether within a species or between species, is a fundamental natural force, which molds the character of species and, in some (e.g., *Homo sapiens*), the character of individuals also. Different species have evolved different ways of surviving in the presence of their competitors, but three basic strategies are usually encountered – the strong specialist, the generalist, and the opportunist. The most obvious strategy, of course, is to become a strong competitor and either drive out or exclude rivals, or seize the disputed resources by aggression, trickery, or other specialized behavior – the "force of arms" strategy. Such strong competitors will tend to be *specialists*, concentrating on particular resources and using specialized weapons and behavior. The strategy of the strong competitor results in a strong negative impact on the rival species, which – in effect – tends to suppress the parameter C. For example, consider the system in Figure 4.6B and imagine that species B develops an advantage over its competitor, so that C_b is depressed towards K_b. The equilibrium point will then shift in favor of species B. If the advantage is great enough then C_b may be reduced below K_b, in which case B will eventually displace its rival from the system (e.g., Figure 4.5C).

Amensalism (–0), where one species has a negative effect on another but is itself unaffected, now reveals itself as a peculiar case where one species has absolutely no competitive ability. Thus, the equilibrium line of the amensal will never intercept the axis of its competitor because it remains at K irrespective of the competitor's density. Interactions with an amensal, B, may stabilize with both species present if $K_b < C_b$, but the amensal will always win when $K_b > C_b$.

The strategy of the *generalist* is to avoid competition by utilizing a variety of resources. The "jack of all trades," with his wide range of alternative resources, can switch quickly from one to the other as the pressures from his competitors change. However, even generalists usually have different preferences from their competitors. This is the first evolutionary step toward specialization, as preference for one resource tends to lead to the selection of characteristics that confer an advantage in obtaining that resource. Hence the classic examples of character displacement amongst closely related species competing for similar resources (e.g., Darwin's finches).

The strategies of the strong competitor and generalist enable them to persist, in the face of their competitors, at population densities close to their saturation levels (e.g., Figures 4.9B and 4.10A). Because of this they often have relatively low maximum rates of increase, and thereby avoid the problem of unstable behavior at equilibrium (see Note 4.7). On the other hand, the strategy of the *opportunist* takes advantage of a high maximum rate of increase and a migrant life style to outwit the opposition. The opportunist uses a "get in and get out" approach as well as the advantage of numbers to obtain a share of the disputed resources. His migrant ways give him the edge in the race to find new resources, and his high reproductive rate enables him to use them up before stronger competitors arrive, or to hold them temporarily through the advantage of numbers.

Opportunistic species are rarely present in any one place for an extended period of time, and so instability at equilibrium created by their high reproductive rates is less of a problem to them. In a way, opportunists are also specialists that are specialized at living in highly variable environments. For example, a variation of the opportunistic way of life is the pioneer who quickly occupies a new environment, often one that has been denuded of life by a catastrophe of nature (or man if you consider him unnatural), and holds it against competitors through the advantage of numbers. Even if the rivals gain entrance it may take a long time to wrest the resources from the well-established pioneers. By that time another natural catastrophe may have paved the way for a repeat performance. The strong competitor and generalist, on the other hand, are more adapted to living in stable environments where their competitive strategies are of greater advantage.

4.4. Predator-Prey Interactions

The subject of trophism, especially predation has fascinated scientists over the ages, and more experimental and theoretical research has been done on predator-prey interactions than on any other single ecological process. This preoccupation is not surprising because man himself has deep predaceous instincts, perhaps being the most efficient predator ever to have appeared on the face of the earth. In addition, predator-prey systems are noted for their interesting and varied dynamic behavior, which is sometimes difficult to interpret and understand.

If we examine the system defined by Figure 4.1 we see that the predator-prey interaction creates an overall negative feedback loop, $N_a \xrightarrow{+} F_b \xrightarrow{+} N_b \xrightarrow{-} F_a \xrightarrow{+} N_a$ where the subscript a holds for prey and b for predator. The loop has a single negative link representing the effect of predator density on the favorability of the prey's environment, which gives it its overall negative feedback effect. This quality should give the interacting system steady-state equilibrium characteristics without the imposition of other regulatory mechanisms, something, which we have not encountered before in our discussion of population interactions. This conclusion is, perhaps, intuitively obvious because predators must be limited by their prey and will, in turn, limit the abundance of their prey. We can see this from Figure 4.1: an increase in the density of A, the prey, raises the favorability of the environment for B, the predator, causing an increase in its rate of increase and population size. The increased predator density then reduces the favorability of the prey's environment, its rate of increase and population density, which in turn reduces the favorability of the predator's environment, and so on.

When we constructed the single-species models in Chapters 2 and 3, we discovered that the most interesting dynamic patterns (steady states, oscillations, cycles) were produced by the action of negative feedback loops. We will find that predator-prey interactions are equally interesting, and that understanding predation will give us a clearer picture of the single-species models.

Let us start by deducing a simple reproduction plane for a prey species. Obviously, the favorability of the prey's environment will be inversely related to the density of the predator, and the equilibrium line may look like that in Figure 4.14A. When the density of the predator is zero, we would expect the prey to equilibrate at its carrying capacity K_a. For each predator added, however, the survival of the prey will be reduced by some amount, which – in line with our previous reasoning – we can call the *marginal cost* of predation W_b (this is represented by the slope of the line). Given that $W_b > 0$, and that the system is linear as in Figure 4.14A, then we

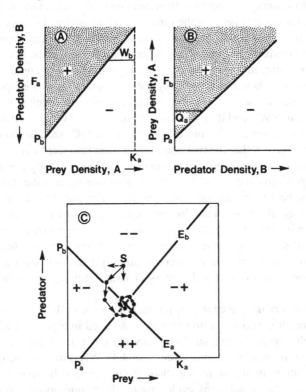

Fig. 4.14 Reproduction planes for a prey species (A), where K_a is the carrying capacity in the absence of predation, W_b is the marginal cost of predation, and P_b is the predator density which drives the prey to extinction; and for a predator (B), where P_a is the minimal prey density needed to sustain a predator population, and Q_a is the marginal benefit of the prey. (C) The superimposed reproduction plane (see Note 4.3 for its construction), showing predator and prey equilibrium lines, E_b and E_a, and a particular dynamic trajectory

will find a particular predator density P_b, which reduces the equilibrium density of the prey to zero (at this point prey are eaten faster than they can reproduce themselves). By this time you may have the feeling that we have done all this before, and you will be right, for the prey equilibrium line is quite similar to that of the competitor. This should not be surprising because predation, like competition, is a cost that must be borne by the species being preyed upon. In a way, we can think of a predator as actually competing with its prey, because both species require the same resource – the energy stored in the prey's body – in order to survive and reproduce.

Let us now turn our attention to the reproduction plane of the predator and, for the present, assume that only one species of prey is eaten; that is, it is a *specific* predator. Naturally, when the prey is absent from the environment the predator will not be able to exist for long because it has no food. In fact we might also suspect that the predator population will only be able to persist after the prey has reached some critical density P_a (Figure 4.14B). Below this level the prey are too sparse and hard to find to sustain the predators. However, above P_a each additional prey added will improve the environment for the predator and, therefore, the slope of the equilibrium line Q_a represents the *marginal benefit* of the prey to the reproduction and survival of the predator. Note that the predator reproduction plane is very similar to that of a cooperator. In a way, the prey is cooperating with its predator by being there for it to eat, although it is certainly not a voluntary form of cooperation.

We can superimpose the two reproduction planes by rotating the predator's clockwise through $90°$, and then the whole thing can be inverted so that the origins are justified to the lower left-hand corner (Figure 4.14C; see also Note 4.3). The first thing we notice is that the two equilibrium lines intersect to create an equilibrium point for most values of their parameters. In fact, the only condition for an equilibrium point to exist is that $K_a > P_a$; in other words, the saturation density of the prey must be larger than the minimum density required to sustain a predator population. This, of course, is an obvious requirement. As we can see, the equilibrium lines run in opposite directions (one runs right-down, the other one runs right-up), in marked contrast to those of cooperators and competitors, which run in the same directions (both run right-up in cooperative systems and both run right-down in competitive systems) and only intersect over fairly narrow ranges of their parameters.

The dynamics of the predator-prey interaction can be evaluated as we have done previously and, if we do this with the system depicted in Figure 4.14C, we obtain a trajectory that spirals in to a stable equilibrium point (the student is encouraged to draw several of these trajectories; see Note 4.3). When we plot the dynamics of the two populations in time series we obtain the trajectories shown in Figure 4.15A. As we can see, the populations cycle around their equilibrium levels, with the predator following the prey, until they reach their steady states after a series of damped cycles.

Although predator-prey systems invariably possess an equilibrium point it needs not necessarily be stable. As we have seen, the system cycles towards equilibrium and this may lead us to suspect that cycles of increasing amplitude may occur under

certain conditions (e.g., Figure 4.15B). In order to investigate the stability of predator-prey interactions we will need to examine the parameters in a little more depth.

First consider the predator reproduction plane under the condition that the prey
population remains constant and above the critical density P_a. The predator population will eventually equilibrate at a characteristic density, provided that the interaction
is stable. However, we might ask: "What determines the density of the predator population living in this constant food supply?" Well, it seems obvious that a predator that
consumes many prey in order to produce a single offspring will not be able to persist
at as high a density as one that requires only a few prey. Hence, the equilibrium density of the predator depends on its efficiency at converting prey into predator offspring. As the equilibrium density is a function of the slope of the line, or the marginal
benefit of the prey Q_a, then this parameter is related to the conversion efficiency of
the predator. In other words, the benefit of a single prey added to the environment is
measured in terms of the predator offspring it can produce and support.

Now let us look at the other predation parameter, the minimal density of prey
needed to support a predator population P_a. It seems reasonable that predators that are
very efficient at hunting and capturing their prey will be able to persist at rather low
prey densities and, therefore, P_a will be relatively small. In contrast, the equilibrium

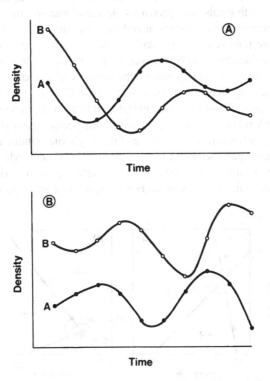

Fig. 4.15 (A) Time-series simulation of the predator (o) and prey (•) populations governed by the
system illustrated in Figure 4.14C, and (B) the same interaction when the slope of the predator
equilibrium line, Q_a, is increased to create an unstable interaction

line for inefficient hunters will intercept the prey axis at much higher densities. Thus, the overall form of the predator equilibrium line, as determined by Q_a and P_a, seems to be related to the efficiency of the predator in seeking out and capturing prey and in converting them into predator offspring (see also Note 4.8).

When we make the predator very efficient, by increasing the slope of the equilibrium line, we will find that the equilibrium point becomes less stable and that eventually we will get cycles of increasing amplitude as illustrated by Figures 4.15B and 4.16A. It should be noted, however, that stability is affected by time delays as well as by the efficiency of the predator. In fact, when there are no time delays in the response of the predator to changes in the density of its prey, then the system will be stable regardless of the predator's efficiency (see also Note 4.9). On the other hand, when the time delay is long we may find unstable interactions involving fairly inefficient predators. The subject of time delays in the predator-prey interaction is extremely important and will occupy our attention later in this chapter.

In the meantime, let us turn our attention to the prey reproduction plane and imagine how the slope of the equilibrium line, or the marginal cost of predation, will change under a constant level of predation. First, the marginal cost of predation should be related to the number of prey that are removed per unit of time, and also to the relief that the survivors obtain by the removal of their competitors; that is, by removing some individuals, the predators decrease intraspecific competition for resources and increase the reproduction and survival rates of the survivors. The rate at which prey are removed by a particular species of predator depends, to a large extent, on their vulnerability to be attacked by that predator. On the other hand, the relief gained by the survivors depends on the intensity of the struggle for resources; that is, on the density-dependent coefficient s (see Chapter 2). Hence, we might expect that the marginal cost of predation is directly related to the vulnerability of the prey and inversely related to the strength of its intraspecific interactions (in Note 4.9 we show that $W_b = v/s$, where v is a measure of the prey's vulnerability to be attacked).

The second prey parameter, P_b, is the density of predators, which can drive the prey population to extinction. Once again, we would expect this critical point to be related to the vulnerability of the prey. However, because intraspecific competition

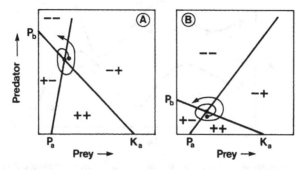

Fig. 4.16 Unstable predator-prey interactions resulting when predators are too efficient (A) or the prey too vulnerable to being attacked (B)

is negligible near to the extinction point, the survivors would not be expected to gain much advantage by the removal of their cohorts. On the other hand, we would expect a prey population with a high maximum individual rate of increase to have a greater chance of persisting in the face of a constant rate of predation than one with a low maximum individual rate of increase. Thus, the extinction threshold, P_b, should be directly related to the maximum individual rate of increase of the prey (R_m in Chapters 2 and 3) and inversely to its vulnerability to attack, v (we show, mathematically, that $P_b = R_m/v$ in Note 4.9).

The effect of increasing the vulnerability of the prey, and hence the marginal cost of predation, is illustrated in Figure 4.16B, when compared with Figure 4.14. Once again we see in this Figure that the predator-prey interaction is destabilized if the prey become too vulnerable to being attacked, given that time delays are present in the system.

In addition to the effect on stability, the slopes of the predator and prey equilibrium lines also determine the relative density of the two species at equilibrium (Figures 4.14 and 4.16). As we would expect, the system equilibrates at high prey and low predator densities when inefficient predators attack prey with good escape or defense mechanisms. Such conditions also favor stable interactions and, as a result, we usually find stable systems when the prey population is much larger than that of the predator. However, we will see later that stable interactions between efficient predators and vulnerable prey are possible if the predators limit their own population size, or if the prey becomes very hard to catch when their density is low.

4.4.1. Nonlinear Predator-Prey Interactions

We have seen that the predator-prey interaction is quite sensitive to the slopes of the respective equilibrium lines. However, there is no reason to suppose that these slopes are always constant so that we get linear equilibrium lines as in Figures 4.14 and 4.16. Some predators, for example, are very efficient at seeking out and capturing their prey, but are inhibited when their populations become very dense because they interfere with each other's hunting activities (e.g., insect parasitoids; see Note 4.10). The equilibrium line for such a predator may be very steep at first, but its slope will decrease in direct relationship to the density of the predator population, and the reproduction plane will look like the one in Figure 4.17A. Even though these predators may be very efficient, their interaction with the prey may be quite stable if their saturation density is not too much greater than the prey's extinction point (Figure 4.17B; cf. Figure 4.16A in which an efficient predator is not inhibited by its own density). We can also see that efficient predators with self-inhibiting interactions can regulate their prey at quite low densities, a result of considerable importance in the biological control of pests.

In contrast, some predators have very inefficient mechanisms for searching out their prey. This is particularly true for pathogenic microorganisms (at least those that are not transported by insect vectors), which reach their hosts by passive transmission through the atmosphere or by contact between infected and uninfected individuals. Although these "predators" are able to survive at low host densities, by entering a dormant state, utilizing other hosts, and so on, their populations will not increase on a particular host until its density gets quite high. However, once this critical density is reached, they can reproduce huge numbers of offspring very quickly so that the slope of their equilibrium lines becomes very steep (Figure 4.17C). Interactions with "predators" of this kind may result in high amplitude cycles, or epidemics, of the pathogen population (Figure 4.17D).

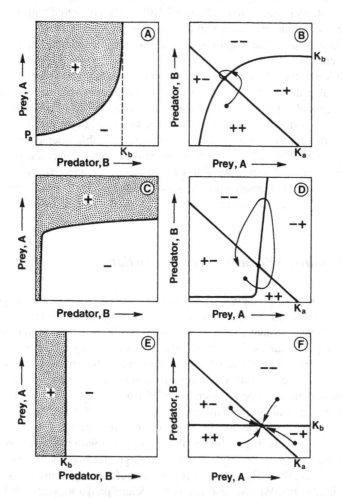

Fig. 4.17 (A) Reproduction plane for an efficient predator which is limited at K_b by intraspecific interactions, and (B) its interaction with a linear prey equilibrium line. (C) Reproduction plane for a predator that responds with great efficiency only after its prey reaches a fairly high density, and (D) its interaction with a linear prey equilibrium line. (E) Reproduction plane for a predator that is limited at K_b by intraspecific interactions and does not respond to prey density, and (F) its interaction with a linear prey equilibrium line

Other predators are relatively independent of the density of a particular prey because they feed on a large number of different species. These general predators are usually limited in numbers by competition with their own kind for hunting territories (e.g., most carnivorous birds and mammals). If they are completely independent of the density of a particular prey, then their equilibrium lines will have zero slope and so will run parallel to the prey axis. In this case the predatory-prey interaction will have a very stable non-cycling equilibrium, provided that the predator's saturation density is below the prey's extinction point (Figure 4.17E,F). Even if the general predator responds numerically to the density of a particular prey species, the equilibrium line will intercept the predator axis at a positive density, or negative P_a, because the predator will be able to exist in the absence of this prey by feeding on other species. However, if the predators are too efficient or the prey too vulnerable, the interaction may be unstable as demonstrated by Figure 4.16A,B.

Of course, we can also visualize various types of prey reproduction planes. For instance, some prey become very vulnerable to being attacked at high population densities because their defense or escape mechanisms are weakened by severe competition amongst themselves. This is particularly true of the vulnerability of organisms to infectious pathogens. In these cases the prey's equilibrium will steeply decline with increasing predator density at high prey densities, but it will decrease only slowly with increasing predator density at low prey densities, and we will obtain a reproduction plane like that in Figure 4.18A. The interaction with an efficient predator will still tend to be cyclic but it will be more stable than the linear case (Figure 4.18B; cf. Figure 4.16A). Because infectious pathogens often have reproduction planes such as that illustrated by Figure 4.17C, an interesting exercise is to superimpose this plane on that shown in Figure 4.18A (see Exercise 5). This interaction will usually be characterized by population cycles, or epidemics of the pathogen, a result that is in line with our observations of real-life host-pathogen interactions.

A similar kind of prey reproduction plane will result if the environment contains a limited number of good hiding places, or refuges from predation. In this case the vulnerability of the average individual will be small when the population is not very dense, because most individuals will be able to find a hiding place. However, when population density increases and all the refuges become occupied, those unfortunates without hiding places will be very vulnerable to predation. In Figure 4.18C a reproduction plane is drawn for a prey that is completely inviolate until all the refuges are filled up, after which it becomes highly vulnerable to its predators. To make things more interesting we have also assumed that the prey is limited by other environmental resources when predator density is very low. The interaction of this prey population with an efficient predator is stabilized by the refuge and the self-limitation of the prey, but we may observe continuous population cycles (Figure 4.18D).

Up to this point we have worked under the assumption that the predator actually kills its prey or, at least, severely impairs its reproduction and survival. However, many predators do not need to kill their prey in order to feed on them (e.g., sap-sucking and blood-sucking animals, fruit-eaters, browsers, and many parasitic animals and plants). In these cases the prey can withstand extremely high densities of the "predator" with

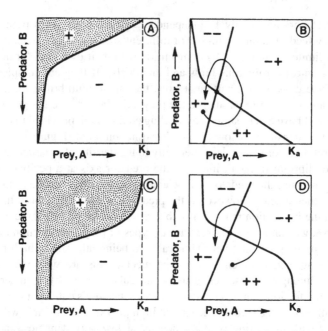

Fig. 4.18 (A) Reproduction plane for a prey that is more vulnerable to its predator at high population densities, and (B) its interaction with an efficient predator. (C) Reproduction plane for a prey that has a fixed number of safe refuges from predation and is also self-limited near K_a, and (D) its interaction with an efficient predator

only minimal impact on its survival and reproduction, and the prey equilibrium line will have a very shallow slope. In effect, the vulnerability of the prey has been reduced by the specialized feeding behavior of the "predator," and this will result in very stable interactions with both predator and prey at high densities close to their saturation levels (the student is encouraged to demonstrate this graphically).

When the feeding of a predator has little or no effect on the reproduction and survival of its prey, then very high populations of both species can coexist. However, an interesting phenomenon arises from this kind of interaction: Because the prey is not destroyed, room is available for another predator to enter the system. Thus, it is no surprise to find that aphids, mosquitoes, and the like often carry pathogenic microorganisms, which use the primary predator as a vector to reach the same host. Obviously this would not be a very successful strategy if the primary predator destroyed its prey.

Two important generalizations have emerged from our examination of various predator-prey reproduction planes. First, stable interactions are likely to occur, even if the predators are very efficient, when the predators are limited to densities below the level where they force the prey population to extinction. Second, stability can also be increased if the prey population is less vulnerable, or better still, immune to attack when its density is low. Both these conclusions apply equally to specific or general predators and, although they may seem intuitively obvious, they are very important to the population manager.

4.4.2. Predator Functional Responses

Until now we have been concerned with changes in the numbers of predators as their populations respond to the density of their prey. This is usually referred to as the *numerical response* of the predator. We have seen that the numerical interaction between predators and their prey is frequently cyclical in nature because of delays in the reproductive response of the predator (Figures 4.14 and 4.15). What happens, in effect, is that each predator is able to capture more food as the prey becomes more abundant, and this food is converted, after a delay, into more predators. However, we can see from this chain of events that two processes are involved in the overall predator response. First, each predator reacts to the density of the prey by eating more or less of them, and second, the prey that are eaten are eventually transformed into predator offspring. The primary feeding response of the individual predator is usually termed its *functional response* in order to differentiate it from the reproductive numerical response (see Note 4.11).

In contrast to the numerical response, the functional response of the predator is immediate because, given more food, the predator immediately eats more. This distinction is important because, as we have learned, fast-acting negative feedback loops tend to create more stable equilibria (Chapters 2 and 3).

Throughout this book we have actually recognized the functional response of predators as an intrinsic part of the prey's density-dependent regulating mechanism. For instance, in Chapter 3 we argued that competition for food might weaken certain individuals so that they become more vulnerable to predation. In addition, as prey density increases, hiding places or escape routes will become overcrowded so that some individuals will be exposed to their predators and become easier to catch. In effect, therefore, competition between the prey for food and space affects their vulnerability to attack and, indirectly, the rate of feeding by predators as expressed by their functional responses. From this line of reasoning we can argue that the functional response of predators should be considered an intimate component of the prey's reproduction plane. This line of reasoning is also consistent with our conceptual model of the single-species population (Figure 3.9) because, if you remember, we considered the density-dependent regulating function to be composed of all those factors that act rapidly in response to population density, while delayed feedback operated when the population affected the properties of its gene pool or its environment. As predator functional responses are immediate and do not involve a change in the properties of the environment, because the number of predators present does not change, then it seems reasonable to include them in the population regulating component of the system.

If we accept the argument that predator functional responses can be included in the prey's reproduction plane, then we need to determine if different kinds of responses affect the form of this plane differently. There are three basic types of functional responses (Figure 4.19A), but all of them have the same fundamental property, in that the number of prey eaten per predator per unit time increases with prey density until the predators become satiated. At this point the response levels off, so that the number of prey consumed per predator remains the same irrespective of prey density. However, the responses have some subtle, and very important differences.

Let us take a closer look at these three basic functional responses. In the first, the number of prey attacked increases linearly with prey density and then suddenly stops when the predators are satiated. This type of response seems to be rather rare in nature, but may be characteristic of some filter feeders, which spend little or no time pausing after each prey is captured; that is, they do not need to stop hunting in order to kill and devour their prey. On the other hand, the type II response is typical of predators that pause after each prey is captured and, therefore, their rate of attack declines as the density of their prey increases. This type of response seems to be typical of many invertebrate predators, but it should be noted, however, that most of the data come from laboratory experiments. Type III functional responses are characteristic of predators that attack their prey at an increasing rate as prey density rises, but then the rate of attack declines as handling time becomes a factor in determining how fast prey can be caught. It is generally thought that type III responses are typical of general predators, particularly vertebrates, which switch their attack to a particular prey species when it becomes more abundant; that is, they learn to look for, or develop a "searching image" of, the more abundant species in their repertoire of prey. However, these responses have also been found in some insect

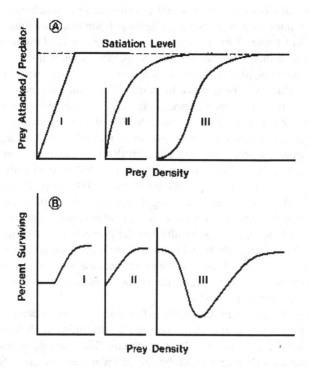

Fig. 4.19 (A) The three basic forms of the predator functional response, and (B) the corresponding percentage survival of a prey population subjected to predation by the three types of functional responses (see Note 4.12 for the method of transformation)

parasitoids, and they may be more common in nature than was previously supposed (see Note 4.13).

The impact of the three types of functional responses on the prey population can be seen by calculating the percentage survival of the prey when subjected to the three kinds of predation (Figure 4.19B and see the corresponding file in EXCEL in the disk that comes with this book and Note 4.12 for computational procedures). We can see from these graphs that type I and II responses allow an increasing proportion of their prey to survive as the density of the prey rises. This creates a positive feedback effect and, consequently, these responses cannot act as density-dependent regulating factors. On the other hand, the S-shaped type III response causes prey survival to decrease at first but then, as handling time and satiation become important factors, survival increases with prey density in a similar manner to the type II response. The negative feedback that occurs at the lower prey densities suggests that predators with type III responses can act as regulating agents in the lower ranges of prey density. However, once the prey population attains higher densities this negative feedback regulating effect is disengaged. As we will see later, this is an extremely significant result.

Let us now try to see how the type III functional response may affect the reproduction plane of a prey population. Suppose a certain percentage survival from predation is necessary for the individual rate of increase to be zero. In other words, mortality from predation and all other factors in the environment exactly balances the number of prey born in a given period of time. The $R = 0$ line will cross the survival curve at two points, if at all (Figure 4.20A). If we draw survival curves for several different predator densities, we will obtain a set of equilibrium points, which can then be used to construct the reproduction plane (Figure 4.20B). If the prey population is regulated by competition for resources at high densities, then the equilibrium line will reach its maximum at the carrying capacity, K_a, as shown in this figure (a similar reproduction plane is derived mathematically in Note 4.14). This reproduction plane may look familiar to those who remember the W-shaped curves of Chapter 3 (e.g., Figure 3.16). It is, in actuality, identical except that the low-density cooperative interaction is missing. As we know, the midsection of the equilibrium line (the broken line in Figure 4.20B) represents a set of unstable equilibria, which we originally interpreted as being due to cooperative interactions. In a sense, a form of unconscious or *de facto* cooperation is occurring in the predator-prey interaction because, as prey density rises above a certain level, the addition of an individual increases the survival of its cohorts (Figure 4.20A).

We can now superimpose predator and prey reproduction planes as we did previously. Because many predators with type III responses are vertebrate generalists, who are limited by things other than the abundance of a particular prey, then their equilibrium states will appear as a horizontal line across the superimposed plane (Figure 4.21A,B). Interactions with these predators may create one or two stable equilibria, depending on the density of the predator population. This result is extremely important to those involved in the management of predator-prey populations. For instance, if the prey is a useful resource, then we can see that the quantity of the resource can be significantly increased by decreasing the predator's density

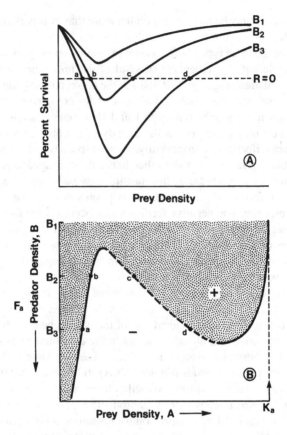

Fig. 4.20 (A) Percentage survival of a prey population when subjected to three different densities of predators, B's, when the predator has a sigmoid type III functional response; the survival requirement to maintain an equilibrium prey population is shown as a broken line $R = 0$. (B) The N-shaped prey equilibrium line produced when the points a, b, c, d are transposed directly from the graph above (see also Note 4.14)

from K_b' to K_b (Figure 4.21B). Conversely, if the prey is a pest, then it can be regulated permanently at a very low density by increasing the carrying capacity for the predator population (e.g., providing nesting boxes for hole-nesting birds).

Important management implications also arise where predator-prey interactions create two potentially stable equilibria (Figure 4.21A). Here the system will move toward one or the other of its equilibrium states depending upon which side of the unstable threshold (the broken line) it starts. However, even if the system is at a particular equilibrium point, it can be moved into the domain of the other by outside disturbances. For example, suppose the prey is a useful resource, which is being harvested for food and that, prior to harvesting, the population was at the higher equilibrium near to its carrying capacity, K_a (Figure 4.21A). Provided that the harvest does not reduce the population below the unstable threshold, then the population will always tend to return to its upper equilibrium. However, if the population is overharvested, or if the harvest plus a natural catastrophe reduces it below the unstable threshold, then the population will move to its lower equilibrium and remain there even if harvesting is discontinued (see, for example, the salmon problem illustrated in Figure 3.6). The only way that the stock can be re-established at its previous level of abundance is through hatchery operations, which raise the density above the critical threshold, or by a temporary predator control program (note that the predators need only be controlled for a short period of time to allow the prey population to rise above the critical threshold).

On the other hand, if the prey is a pest that is regulated at the lower equilibrium by its predators, then it may be displaced into the domain of the upper equilibrium by an outside disturbance. Such a disturbance may take the form of a flight of insect pests from your neighbor's fields, which raises the density of the resident population above the unstable threshold, or it may be the application of a pesticide for another pest species, which kills off the predators. Whatever the reason, the pest population will then grow toward its upper equilibrium where it will cause considerably more damage to the crop.

If type III functional responses are found in specific predators, or in general predators which respond numerically to the density of their prey, then we can obtain a variety of dynamic behaviors depending on where the predator equilibrium line

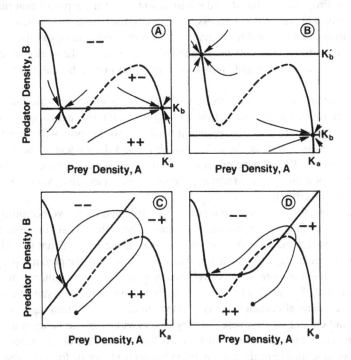

Fig. 4.21 Several predator reproduction planes superimposed on the N-shaped prey equilibrium line; (A, B) general predators with no numerical response to prey density, (C, D) general predators with different kinds of numerical responses to prey density – K_a and K_b are the carrying capacities of prey and predator, respectively

crosses that of its prey. For example, we may observe cyclic outbreaks of the prey population whenever it transcends the unstable equilibrium (Figure 4.21C,D). Population systems that have a single stable equilibrium but that can exhibit radical disruption following a minor disturbance are called *metastable*. The latter case may be more usual with general predators. Here the predators remain more or less unaffected numerically when the prey is relatively sparse but, when the prey becomes very abundant relative to the food supply in other areas, predators may migrate in large numbers to this abundant food source (this will be discussed in more detail in the next chapter). These results are significant because they help to explain why some populations remain at rather low densities for long periods of time but then go through eruptive population cycles, outbreaks, or epidemics.

4.4.3. Predation in Variable Environments

We have analyzed the interactions between predator and prey populations under the assumption that all the other properties of the shared environment remain constant with time. It is now time to look at the effects of variable environments on this interaction. Properties of the shared environment affect the reproduction planes of both predator and prey, but not necessarily in the same way nor to the same degree. Certain environmental conditions may favor the predator and others the prey. The important question is how will environmental changes affect the steady-state densities of predator and prey populations and, more importantly, the stability properties of the interaction?

First considering the prey reproduction plane, we would expect the shared environment to influence the saturation density, K_a, and the extinction point, P_b. Environments that provide more essential resources for the prey will permit higher saturation densities to be attained. We can show, and the student is encouraged to do so, that raising K_a will have rather minor effects on the equilibrium densities but will tend to destabilize the predator-prey interaction (see also Note 4.15 for an example of this effect).

As we know, the prey extinction point, P_b, is affected by the vulnerability of the prey and its maximum per capita rate of increase. Thus, shared environments that are more favorable for the reproduction and survival of the prey, or that provide more hiding places and escape routes, will increase the stability of the predator-prey interaction. As organisms are usually well adapted to their environments, we might expect that both K_a and P_b will be higher in more favorable environments, and that their counteracting effects on stability will tend to maintain the *status quo*. However, we may find environments where the prey is very vulnerable to being attacked and yet which permit high saturation densities. Such conditions are often created in laboratory experiments where the prey is provided with plenty of food but few hiding places or escape routes. It should not surprise us, therefore, that most of these experimental predator-prey systems prove to be unstable (Figure 4.22 and Note 4.16).

In a similar manner, variations in the shared environment may modify the parameter P_a of specific predators and, if their populations are limited by resources other than the prey, their saturation densities, K_b, as well. Naturally, the efficiency of a predator as it searches for prey is affected by physical impediments to its rate of movement or to its tracking ability – barriers, rain, snow, wind, temperature, etc. – whereas the saturation density is determined by such things as nesting places. From our previous analyses we can conclude that any environmental change that favors a specific predator, or for that matter a general predator that responds numerically to the density of its prey, will tend to destabilize the predator-prey interaction. Once again we will find such conditions in many laboratory ecosystems, which turn out to be unstable (see Note 4.16). However, when these artificial ecosystems are changed to make things more difficult for the predator, or to provide hiding places or escape routes for the prey, then they become more stable (Figure 4.22).

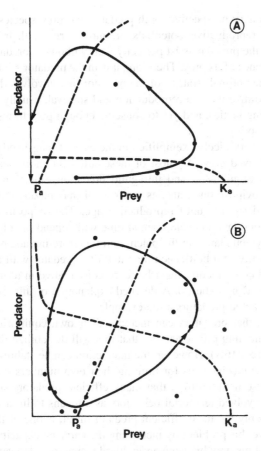

Fig. 4.22 Reconstruction of C. B. Huffaker's experiments with a predatory mite (see Note 4.16 for reference): (A) Trajectory leading to the extinction of both prey and predator in a simple environment, and (B) cyclic trajectory when prey and predator coexist in a more complicated environment containing barriers to the predator and escape routes for the prey. The broken lines are rough guesses at the prey and predator equilibrium lines

Finally, we have seen in Figure 4.21A,B that changes in the saturation densities of general predators with S-shaped functional responses may radically alter the equilibrium conditions of the predator-prey interaction. In environments that are unfavorable for the predator we may find a stable equilibrium at high prey and low predator density. As the environment for the predator improves, two stable positions may be created; then, in very favorable environments, we may again find a single stable equilibrium, but this time at low prey and high predator density. As we mentioned before, this leads to some important inferences concerning the management of predator-prey systems and their environments.

4.4.4. Predator and Prey Strategies

From an evolutionary perspective, both predator and prey species strive to maximize their own reproductive potentials or, more strictly, their genetic fitness. However, whilst the prey can exist perfectly well without its predator, the predator *requires* the presence of its prey. Therefore, it is in the predator's interest to practice *conservation*. The optimal strategy of the predator then involves the counteracting pressures to maximize its own reproduction and survival, usually through the use of efficient hunting tactics, and yet to conserve enough prey to replenish the food supply for its offspring.

This has been very nicely exemplified in the case of long-lived insect predators feeding on short-lived prey. As most of these suffer an enormous egg and larval mortality due to cannibalism and intraguild predation, selection acts mainly on optimizing their oviposition strategies in terms of maximizing the likelihood that the offspring will survive until reproductive age. The oviposition strategy of a predator with a long larval developmental time will depend on a longer projection of the future prey abundance in the patch, will therefore include more bottlenecks or higher probability of a bottleneck than that of a predator with a short developmental time, and consequently must be more conservative in terms of preserving their prey (the GTR hypothesis). A detailed explanation of this theory is given in Note 4.17, using aphid predators as an example.

We have seen that predators can increase their own equilibrium densities by improving their hunting efficiency, but that this will destabilize the predator-prey interaction (Figure 4.16A). However, the interaction can be stabilized if the predators have effective mechanisms for limiting their own numbers (Figure 4.17B). It should not surprise us, therefore, that many efficient predators such as the large vertebrates have evolved territorial behaviors as a means to limit their population sizes. Man, probably the most efficient predator of all, is one of the few who has attempted to solve this problem by increasing the carrying capacity and individual rate of increase of his prey by using agricultural technology. However, our analysis indicates that, although this may permit higher equilibrium densities to be attained, the stability of the interaction may remain unaffected or may even be lowered. It is unlikely that many animals, with the doubtful exception of man, have the intellect to rationalize the critical importance of a stable predator-prey interaction. Thus, most predators have had to learn this lesson in the unforgiving arena of evolution. In the past, efficient predators that lacked the genetic "sense" to limit their own numbers must have gone extinct in droves.

Stable predator-prey interactions can also be created if efficient predators limit the impact that they have on their prey. Hence, those predators that do not seriously debilitate their prey can attain very high equilibrium population densities. From this point of view parasitism emerges as a highly effective strategy, at least in those parasites that do not seriously harm their hosts.

In an immediate sense, evolutionary pressures will always tend to select the more efficient predators from the population, because they will succeed in capturing more prey and in producing more offspring. However, the long-term fitness of the species may be lowered by this trend unless selection also favors conservative tendencies; for example, territorial or other self-limiting behaviors or reduced impact on the prey. It is much more difficult to see how natural selection favors these traits without evoking the concepts of group selection (see Note 4.18). In this sense selection operates on the group, or population, rather than on the individual. Those populations that are poorly adapted to their environment or that create unstable conditions are much more likely to become extinct. Thus, predator populations that evolve highly efficient hunting behaviors, but that fail to evolve methods for limiting their own numbers or their impact on their prey, may flourish for a time but are much more likely to crash to extinction.

In contrast to the predator, the strategy of the prey is simply to maximize its own reproductive potential (here we are strictly avoiding the question of the prey conserving its own food resources and are restricting ourselves purely to its role as the prey), in other words, to avoid being eaten by its predators. There are a number of tactics that a prey can evolve to achieve these ends: locomotory and sensory systems may be adapted for sensing and fleeing predators, cryptic habits and camouflage permit some organisms to hide, while sessile organisms usually have well-developed defensive systems that may involve distasteful or toxic chemicals and/or physical structures such as shells or spines. Most organisms also possess internal defensive mechanisms for dealing with parasitic invasions – phytoalexins, antibodies, leukocytes, and the like.

Perhaps the most interesting defensive tactics have been evolved by organisms that are attacked by predators with learning abilities. Some of these have an unpleasant taste but, instead of being camouflaged, they go out of their way to be noticed with bright, distinctive colors and patterns. The intelligent predator, having tasted one of them, carefully avoids its brightly marked and unpleasant brethren. Even more intriguing is the mimic who, being quite tasty himself, dons the attire of his distasteful associates and thereby fools the predator. As you can imagine, there are many mimics of wasps, bees, ants, and other potent species. The mimic, however, has a problem, which is not unlike that of the predator: if he becomes too numerous, so that the predator encounters his kind too frequently, then the learning may be reversed. The mimic must practice population control if its strategy is to work. (An interesting exercise is to evaluate the interaction between mimic and model. The reproduction planes and equilibrium solutions will be similar for predator and prey.)

In the continuous struggle between predator and prey, the latter usually has the evolutionary advantage. The pressures of predation will select those prey individuals that are better able to escape or defend themselves. This may put pressure on the predators to evolve more efficient hunting behavior. However, unless the prey's tactics are very powerful, this needs not happen. Although the prey may be more difficult to catch at first, soon there will be more of them making them easier to capture again. Thus, fairly modest changes in the prey's defense or escape tactics may place little or no selective pressure on the predator. In contrast, if the predator evolves more efficient hunting techniques, then the prey population will be depressed, making them more difficult to find and the system will become more and more unstable. In fact, it may sometimes be advantageous for the predator to evolve less efficient hunting behaviors. This will result in increased prey density, greater stability, and may even result in a higher predator equilibrium density. For example, the females of some insect "predators" of forest trees have lost their ability to fly, thereby lowering their searching efficiency (e.g., the Douglas-fir tussock moth and the gypsy moth). Thus, the co-evolution of predators and their prey is controlled by a complex pattern of interdependencies, which – in the long run – tend to produce a finely tuned and balanced interaction. We will have much more to say on this subject when we consider spatial interactions in the next chapter, and in the last chapter we will see that predator-prey interactions may even be viewed as being mutually beneficial to both predator and prey species.

4.5. Chapter Summary

In this chapter we have examined interactions between two species cohabiting the same geographic region by considering them as separate subsystems interacting through their shared environment. These interactions were evaluated by superimposing the reproduction plane of one species on that of the other. The evaluation criteria were the densities of each species at equilibrium and the relative stability of the interaction. The main points are briefly summarized below:

1. Interactions were classified according to the effect of each species on the favorability of the other's environment as cooperative (++ or +0), competitive (– – or –0), and predator-prey (+ –).
2. Cooperative interactions form an unstable positive feedback loop, which results in indefinite growth of both species unless one or the other is limited by other environmental factors.
3. Competitive interactions also form an unstable positive feedback loop, which can only be stabilized if one or both species are limited by other environmental factors.
4. Permanent coexistence of competing populations is only possible if the negative effect of each species on the members of its own population is greater than its

effect on its competitor's; that is, $K_a < C_a$ and $K_b < C_b$. In all other cases one species will win the contest and the other will eventually be excluded.

5. The equilibrium densities of the competing populations are determined by the intensity of the interaction, higher densities being attained when the competitors have different preferences or utilize alternative resources.

6. Under certain conditions (i.e., when $K_a > C_a$ and $K_b < C_b$) the interaction between two competitors may create a double equilibrium system in which species A exists alone or both species coexist, with species B at a much higher density.

7. Environmental changes may be very important in determining whether competing species coexist or one or the other dominates. In particular, when one species reduces the favorability of its own environment while having no effect or even improving that of its rival, then it may eventually be replaced, giving rise to ecological succession.

8. Strong competitors are often specialists that utilize force to wrest resources from their rivals or generalists with a wide range of alternative resources to choose from. These species usually have rather low maximum per capita rates of increase and do best in stable environments.

9. Weak competitors often exist in nature by being opportunists, seizing and utilizing resources rapidly before the stronger rivals arrive on the spot. For this reason they are usually highly mobile organisms with high maximum rates of increase and are best adapted to more unstable or harsh environments. Pioneer species also use the advantage of numbers to temporarily hold resources from their stronger rivals.

10. Predator-prey interactions form a negative feedback loop whose stability depends on the properties of both predator and prey. Unstable interactions occur when highly efficient predators attack very vulnerable prey. However, this interaction can be stabilized if the predators are limited by other environmental or behavioral mechanisms, if their impact on the prey is minimized, or if the prey has refuges from predation.

11. The predator-prey interaction is often of a cyclic nature. Continuous cycles may sometimes be observed if the system is on the borderline of instability, if time delays are present in the feedback loop, if predators are inefficient at low prey densities but efficient at high densities, or if prey attacked by efficient predators have refuges from predation.

12. The steady-state densities of the two species are determined by the efficiency, impact, and self-regulatory characteristics of the predator and the vulnerability of the prey. Stable interactions will usually be found when the system equilibrates at high prey and low predator densities, unless the predator limits its own numbers, or the prey have safe refuges from predation.

13. General predators which switch their attack from one prey species to another often have S-shaped functional responses. Such functional responses create complex interactions, which may have one or two stable equilibria, or exhibit cyclic behavior, depending on the numerical response of the predator population.

14. Environmental changes that affect the vulnerability of the prey, the efficiency of the predator, or the saturation density of either species, may have severe effects on the stability of the predator-prey interaction.

15. The optimal strategy for the predator involves the conflicting goals of maximizing its reproductive potential and yet conserving prey for future generations. The most effective strategy combines maximum efficiency with self-limiting behavior such as territoriality.

16. The strategy of the prey is simply to maximize its own reproductive potential by escaping its predators using mobility, hiding places, camouflage, defensive weapons, or mimicry. Evolutionarily, the prey is usually one jump ahead of its predator.

Exercises

4.1. Analyze the model

$$A_t = A_{t-1} + (R_{ma} - s_a A_{t-1} - c_b B_{t-1})A_{t-1}$$
$$B_t = B_{t-1} + (R_{mb} - s_b B_{t-1} - c_a A_{t-1})B_{t-1}$$

which expresses the interaction between two competing populations when the repressive effect of each species on its own numbers and on those of its competitor are linearly related to its density (see Note 4.4).

A. What do the parameters s and c represent?

B. Draw the reproduction planes for two competing species when $R_{ma} = R_{mb} = 2$, $s_a = s_b = 0.002$, and (i) $c_a = 0.001$, $c_b = 0.001$; (ii) $c_a = 0.001$, $c_b = 0.003$; and (iii) $c_a = 0.003$, $c_b = 0.003$. Evaluate the stability properties of these interactions by drawing trajectories on the superimposed reproduction plane. What conditions are necessary for these populations to coexist?

C. Suppose we have the system in B(i) above and the environment becomes less favorable for species B so that $R_{mb} = 1$. Draw the superimposed reproduction plane for this interaction and evaluate its stability graphically.

D. Using the disk that comes with this book, run the program for two-species competition in EXCEL. Repeat the exercise above using numerical simulations. What additional information did you obtain from the numerical solutions?

4.2. Explain, using your knowledge of competition theory, how ecological succession may occur.

4.3. Explain how opportunistic species are able to persist in the face of superior competitors.

4.4. Analyze the model

$$A_t = A_{t-1} + (R_{ma} - s_a A_{t-1} - vB_{t-1})A_{t-1}$$
$$B_t = B_{t-1} + R_{mb}(1 - B_{t-1}/eA_{t-1})B_{t-1}$$

which describes the interaction between a prey, A, and its predator, B, under the assumption of linearity (see Note 4.9 for details).

A. What do the parameters v and e represent?

B. Draw the reproduction planes for each species and superimpose them (equations for computing the equilibrium lines can be found in Note 4.9) under the conditions that $R_{ma} = 2$, $R_{mb} = 1$, $s_a = 0.002$, $v = 0.01$, $e = 0.2$. Identify the equilibrium point (A*, B*) you can also find this point mathematically by solving the equilibrium system (see Note 4.9). Perform a steady-state analysis to determine the stability of this system; in other words solve the model numerically when it is displaced from equilibrium (e.g., $A_0 = A* - 100$, $B_0 = B* - 40$.). Use the disk that comes with this book and the program for predator-prey interactions in EXCEL. Plot the trajectory on the superimposed reproduction plane and in time series.

C. Repeat the above analysis with all the parameters the same except let $e = 0.5$. Calculate the equilibrium point and stability properties.

D. Repeat the analysis with $e = 0.5$ and $v = 0.005$ and evaluate the equilibrium point and stability properties.

4.5. An insect is infected and killed by a polyhedrosis virus. However, the insect only becomes vulnerable, or the virus becomes virulent, at very high population density when food shortage causes physiological stress amongst the crowded insects. In addition, the transmission of the virus from infected to healthy individuals is facilitated when the density of the insect becomes high. Draw a superimposed reproduction plane for this "predator-prey" system and evaluate and describe its dynamics.

4.6. In Note 4.14 we showed how an N-shaped prey equilibrium line can be calculated from an equation which includes a type III predator functional response. This is done by finding the predator densities required to maintain particular prey equilibria. Use this method to construct a prey reproduction plane when the prey's maximum individual rate of increase $R_m = 2$, its carrying capacity $K = 200$, its density which begins to saturate the predator $A_i = 20$, the maximum predator attack rate $\varepsilon = 2$, and the predator switching coefficient $n = 2$.

A. Suppose we have an area where this general predator is limited to 20 individuals by a shortage of nesting places. What would be the equilibrium density of the prey?

B. What would happen if we artificially increased the number of nesting sites so that 40 predators could now live in this area?

C. Suppose an environmental catastrophe then killed off 80% of the prey population but had no effect on the predators. What equilibrium density would the prey attain after this catastrophe?

D. If the prey is a pest, how many predators would have to be supported in the area to ensure that the pest population always remained at a low density?

Notes

4.1. A number of classifications have been proposed for interactions between two species, resulting in a sometimes confusing proliferation of terms. Eugene Odum, in his classic text *Fundamentals of Ecology* (p. 211 in the 3rd ed.; W. B. Saunders Co., Philadelphia, 1971), identifies nine kinds of interactions by splitting competition into direct and indirect types, separating predation from parasitism, and dividing symbiosis into obligatory and nonobligatory forms. However, in keeping with the nature of this present book, we have tended to lump together rather than to split apart in an attempt to retain an elemental simplicity. We hope to see that particular types of interactions are the evolutionary result of species interacting in the three basic ways.

4.2. *Linear Cooperation Model.* Consider the linear density-dependent model of Chapter 2, equation (2.6), which describes the dynamics of a single species,

$$A_t = A_{t-1} + (R_{ma} - s_a A_{t-1})A_{t-1},$$

where A_t is the density of species A at time t, R_{ma} is its maximum per capita rate of increase, and s_a is the repressive effect of each individual on the rate of increase of its cohorts. If a cooperator is present in the environment, then each individual of this species, B, will have a positive effect, p_b, on the rate of increase of A. Thus, we can rewrite the equation as

$$A_t = A_{t-1} + (R_{ma} - s_a A_{t-1} + p_b B_{t-1})A_{t-1}$$

and likewise

$$B_t = B_{t-1} + (R_{mb} - s_b B_{t-1} + p_a A_{t-1})B_{t-1}$$

Now species A will be at equilibrium A^* when

$$R_{ma} = s_a A^* + p_b B = 0$$

and the equilibrium line for species A is defined by

$$A^* = \frac{R_{ma}}{s_a} + \frac{p_b B}{s_a}$$

or, substituting K_a, the saturation density of A, for R_{ma}/s_a (Chapter 2), we get

$$A^* = K_a + p_b B / s_a$$

and likewise for species B

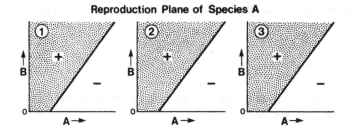

Reproduction Plane of Species A

Reproduction Plane of Species B

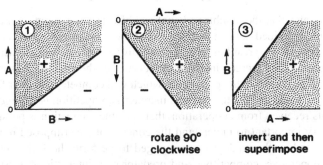

rotate 90°
clockwise

invert and then
superimpose

Fig. 4.23 Transformations of reproduction planes

$$B^* = K_b + p_a A / s_b .$$

We can see that the slope of the line p/s, or the marginal benefit of cooperation Q in the text, is defined as the total beneficial effect of each cooperator divided by the repressive effect of individuals of the same species.

Now the interacting system will be in equilibrium when

$$A^* = K_a + p_b B^* / s_a$$

and

$$B^* = K_b + p_a A^* / s_b .$$

Substituting for B^* in the first equation we get

$$A^* = K_a + \frac{p_b}{s_b}(K_b + p_a A^* / s_b) ,$$

from which

$$A^* = Z /(1 - p_b p_a / s_a s_b) ,$$

where $Z = K_a + p_b K_b / s_a ,$

From this we can see that A^* takes a positive value if, and only if

$$\frac{p_b p_a}{s_a s_b} < 1,$$

which is our criterion for stable coexistence $Q_a Q_b < 1$ in the text. We can also write this criterion as

$$p_b p_a < s_a s_b,$$

which states that the combined beneficial effects of cooperation must be less than the combined self-limiting effects if the populations are to coexist as a stable system. This result is, perhaps, intuitively obvious because when we add a cooperator, and thereby increase the survival of the other species, new members will be added to that population which will compete with their fellows for food and space. If the effect of this increased competition is stronger than the benefits received from cooperation, then a stable interaction is possible.

4.3. It is very important to understand the concept of superimposed reproduction planes, and so the student is encouraged to perform the following exercise with cooperative, competitive, and predator-prey interactions: First draw the equilibrium lines for each species on separate sheets of paper, or better still, clear acetate. Then manipulate one or both sheets until species A occupies the abscissa and species B the ordinate and the origin $A = B = 0$ is in the lower left hand corner. These manipulations are illustrated for the cooperative interaction in Figure 4.23. The reproduction planes can now be superimposed by drawing B's on top of A's, and labeling each equilibrium line, their intercepts, and their zones of population growth and decline.

It is also important for the student to compute several population trajectories on each superimposed plane. To do this start at any point (A,B) in the graph and put arrows for the expected direction each species will move; e.g., if A's sign is positive at this point it will move horizontally to the right and if B's is negative it will move vertically downwards. The distance moved by each species will depend on their positions relative to their respective zero axes and to their equilibrium lines; i.e., population change over the time increment will be smallest close to these lines and greatest in between (see Chapter 3). The distance and direction moved on the reproduction plane will be the resultant of these two vectors (see Figure 4.2C). For purposes of simplicity it is best to assume that the approach to equilibrium is asymptotic. However, we should remain aware that the stability qualities of each system are governed by the slope of the reproduction curve in the immediate vicinity of the equilibrium line as well as time delays in the negative feedback loops. These qualities have been suppressed in our simplified two-dimensional graphical model. However, the rules of feedback specify that if either species is unstable by itself, or exhibits cyclic dynamics, then this effect will be transferred to the two species interaction (see Chapter 6).

4.4. *Linear Competition Models.* If we return to the density-dependent model of Chapter 2 we have equation (2.6), which describes the dynamics of a single population system. This can be written

$$A_t = A_{t-1} + (R_{ma} - s_a A_{t-1})A_{t-1},$$

where A_t is the density of species A at time t, R_{ma} is its maximum per capita rate of increase, and s_a is the repressive effect of each individual on the rate of increase of its cohorts. Now if a competing species, B, is present, then each individual of its population will have a repressive effect, c_b, on the per capita rate of increase of the members of population A. Thus, the total repressive effect on A will be $s_a A + c_b B$, and the equations for the interacting population system are

$$A_t = A_{t-1} + (R_{ma} - s_a A_{t-1} - c_b B_{t-1})A_{t-1}$$
$$B_t = B_{t-1} + (R_{mb} - s_b B_{t-1} - c_a A_{t-1})B_{t-1}$$

Now species A will be in equilibrium, A^*, in the presence of B, whenever

$$R_{ma} - s_a A^* - c_b B = 0$$

and, therefore, the equilibrium line for A is defined by

$$A^* = \frac{R_{ma}}{s_a} - \frac{c_b B}{s_a}$$

and we see that the marginal cost of competition, or the slope of the line, is $W_b = c_b/s_a$. We can also substitute K_a for R_{ma}/s_a

$$A^* = K_a - c_b B / s_a.$$

It is evident that when B is zero then $A^* = K_a$, and when A^* is zero then $B = K_a s_a / c_b$. The density of species B which reduced A's equilibrium to zero was called C_b in the text and, therefore, $C_b = K_a s_a / c_b$. With values for K, s, and c for each species we can draw their equilibrium lines and evaluate the dynamics on the superimposed reproduction plane, as we did in the text, to show that the conditions for stable coexistence are only met when $K_a < C_a$ and $K_b < C_b$. However, we can also demonstrate this mathematically by solving the equilibrium system

$$A^* = K_a - c_b B^* / s_a$$
$$B^* = K_b - c_a A^* / s_b$$

Substituting for B in the first equation we get

$$A^* = K_a - \frac{c_b}{s_a}(K_b - c_a A^* / s_b),$$

which reduces to

$$A^* = \frac{K_a}{(1 - c_b c_a / s_a s_b)} - \frac{c_b K_b}{s_a (1 - c_b c_a / s_a s_b)} \, .$$

From this equation we can see that a positive equilibrium is possible under two conditions:

(1) When $(1 - c_b c_a / s_b s_a) < 0$, then A^* will be positive provided that

$$\frac{K_a}{(1 - c_b c_a / s_b s_a)} < \frac{c_b K_b}{s_a (1 - c_b c_a / s_b s_a)} \, ,$$

which reduces to

$$K_a < c_b K_b / s_a \, .$$

This can be rearranged to give

$$K_b > s_a K_a / c_b = C_b \, .$$

However, a graphical analysis of this equilibrium point will show it to be unstable (Figure 4.6A).

(2) When $(1 - c_b c_a / s_b s_a) > 0$, then A^* will be positive provided that

$$\frac{K_a}{(1 - c_b c_a / s_b s_a)} > \frac{c_b K_b}{s_a (1 - c_b c_a / s_b s_a)} \, ,$$

which reduces to

$$K_a > c_b K_b / s_a \, .$$

This can be rearranged to give the result obtained in the text

$$K_b < s_a K_a / c_b = C_b$$

and we can show graphically that the equilibrium under these conditions is stable (Figure 4.6B).

If we perform these calculations for B^*, we get exactly the same results as in (1) and (2), only the indices a and b are swapped, as the equations for the competitive systems are symmetrical with respect to a and b. The criteria for a stable coexistence between the two competing species are, therefore,

$$K_a < \frac{K_b S_b}{c_a} = C_a \qquad \text{and} \qquad K_b < \frac{K_a S_a}{c_b} = C_b$$

or, substituting R_m/s for K,

$$\frac{R_{ma}}{s_a} < \frac{R_{mb}}{c_a} \quad \text{and} \quad \frac{R_{mb}}{s_b} < \frac{R_{ma}}{c_b},$$

which can be rearranged to give

$$\frac{s_a}{c_a} > \frac{R_{ma}}{R_{mb}} > \frac{c_b}{s_b}.$$

We can see that when the rate of increase of one species is large relative to the other, then it must have a correspondingly strong self-inhibiting effect or a weak interaction with its competitor if the two species are to coexist. When the two species have similar rates of increase, then coexistence is possible if $s_a > c_a$ and $s_b > c_b$; that is, if the self-inhibiting effects of both species are stronger than their competitive effects on the other species.

The competition equations are more commonly found written in continuous time as the so-called "Lotka-Volterra" (see A. J. Lotka's *Elements of Physical Biology*, Williams and Wilkins, Baltimore, 1925) equations. If we let $p = c_b/s_a$ and $q = c_a/s_b$, then

$$dA/dt = r_{ma}(1-(A+pB)/K_a)A$$
$$dB/dt = r_{mb}(1-(B+qA)/K_b)B$$

For those interested in the detailed experimental analysis of competing systems we suggest reading G. F. Gause's delightful book *The Struggle for Existence*, published by the Williams and Wilkins Company, Baltimore, in 1934.

4.5. M. E. Gilpin and F. J. Ayala [*Proceedings of the National Academy of Science (U.S.A.)*, vol. 70, p. 3590, 1973] have analyzed the interaction between two species of *Drosophila* competing for a fixed quantity of food in culture bottles and found equilibrium systems of the type shown in Figure 4.9C,D. Their model explains the nonlinearities in the intraspecific competitive process. However, an equally tenable argument is that the interspecific interaction is nonlinear or, for that matter, that both processes have nonlinear components.

4.6. An interesting review of the rise and fall of various herring, sardine, anchovy, and pilchard fisheries is given by G. I. Murphy in the book *Fish Population Dynamics* (edited by J. A. Gulland, John Wiley and Sons, New York, 1977). Some of these fisheries have collapsed dramatically under heavy exploitation and – in some instances – the collapse seems permanent. However, there is considerable evidence that other similar species have increased dramatically following the collapse of the original fishery (see Figure 4.11).

As an expatriate Cornishman, AAB is keenly aware of the collapse and virtual extinction of the Cornish pilchard schools, and of the current heavy exploitation of mackerel stocks – perhaps these were the competitors that replaced the pilchards? The lessons from competition theory and past experience seem plain and, yet, little seems to be done to rectify these problems.

4.7. Strong competitors are often referred to, following the ideas of Robert MacArthur and others, as K-strategists. The K-strategy aims at maintaining a high but consistent population close to the saturation density (carrying capacity) and is usually most successful when organisms inhabit rather stable environments. K-strategists usually have low maximum rates of increase and fast-acting regulating mechanisms and, therefore, show high degrees of temporal stability.

Opportunists, on the other hand, are called r-strategists, because they have high maximum rates of increase, inhabit variable or temporary environments, and tend to have low temporal stability. Although, for a number of reasons, we have avoided these terms in this book, the following references are provided for those who wish to pursue this subject: *The Theory of Island Biogeography*, by R. H. MacArthur and E. O. Wilson (Princeton University Press, 1967) and T. R. E. Southwood's contribution in *Theoretical Ecology – Principles and Applications*, edited by R. M. May (W. B. Saunders and Co., Philadelphia, 1976).

4.8. The predator parameters can also be viewed at a more basic physiological level using the approach of Andrew Gutierrez and his co-workers (e.g., see the article by A. P. Gutierrez in the *EPPO Bulletin*, vol. 9, p. 265, 1979). From this perspective we consider predator reproduction and survival to be a function of stored energy, which is supplied by eating prey, and physiological time, or aging. If we set the time scale equal to the life span of the predator, then the energy available for reproduction is $S - D_m$, where S is the energy supply, in terms of prey eaten, and D_m is the energy demand for basic metabolic processes in order to keep the predator alive (D_m can be viewed as the number of prey required to meet the basic metabolic demands of the predator). We can see that the predator will starve if $S < D_m$, while if $S > D_m$ there is surplus energy which can be used for reproduction. Given an energy supply in excess of the basic survival demands, then the number of offspring produced will increase, when S (and consequently the supply/demand ratio, S/D_m) will increase, and it will approach its intrinsic maximum, when S (and consequently the supply/demand ratio, S/D_m) will be very large (Figure 4.24).

We now see that the predator parameter P_a is the prey density at which one predator can just gather enough to supply its basic metabolic needs; i.e., where $S/D_m = 1$. However, the supply obtained from a given density of prey is also dependent on the efficiency of the particular predator in searching out and capturing its prey; i.e., S will be higher for a more efficient predator under equal prey density levels and, therefore, S must be related to the hunting efficiency of the predator.

Following the same line of reasoning we can show that the parameter Q_a, the marginal benefit of the prey, is also related to the supply/demand ratio. Given that $S/D_m > 1$, then the number of offspring produced to replace each dying predator is proportional to $(S - D_m)/D_0$, where D_0 is the number of prey

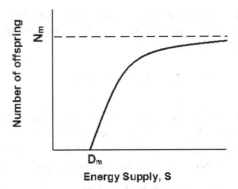

Fig. 4.24 The dependence of the number of offspring produced on energy supply

required to meet the energy demand of producing a single offspring. As more offspring can be produced from a fixed supply of prey when D_0 is small, then the marginal benefit of the prey, Q_a, must be proportional to $1/D_0$.

It is important to recognize the relationship between our simplified population models and the more detailed and realistic physiological models. Although the former are appropriate for evaluating the behavior of generalized population systems, as we are doing in this book, the latter are usually much more powerful for evaluating the behavior of specific population systems. However, as we have seen in this note, both approaches are closely interrelated and the differences are largely in the perspective of the model builder and in his reasons for building the model. We should also realize that both approaches are simplifications of real life and that population systems can also be viewed at even more microcosmic levels of organization; i.e., the organs and their energy demands, the cell, the gene or even at the chemical level (Chapter 1).

4.9. *Linear Predation Models.* If we assume that the prey population grows "logistically" according to equation (2.6), and that the effect of predation on the reproduction and survival of the prey is linearly related to the density of the predators, then we can write the prey equation

$$A_t = A_{t-1} + (R_{ma} - s_a A_{t-1} - vB_{t-1})A_{t-1} ,$$

where A_t is the density of the prey population at time t, B_t is the density of the predator population, and v is the vulnerability coefficient of the prey. In reality v is a rather complex parameter resulting from the interaction between predator attack and prey defense behaviors, which determine the rate of attack, and whether the prey dies or merely has a reduced chance to survive and reproduce when attacked, which determines the debilitating effect, or impact, of an attack. The vulnerability coefficient, therefore, includes those predator and prey attributes that affect the rate of attack and the impact of each attack on the reproduction and survival of the prey.

Turning to the predator equation, let us assume that the carrying capacity of the predator population is determined by the density of its prey so that from text equation (2.6) we get

$$B_t = B_{t-1} + R_{mb}(1 - B_{t-1}/eA_{t-1})B_{t-1},$$

where $eA_{t-1} = K_b$, and e represents the number of predators that can be sustained at equilibrium by each prey, or the efficiency of converting prey into predators. In other words, $e = w_b$, the marginal benefit of the prey. This equation assumes that the predator is perfectly efficient at capturing prey so that $P_a = 0$.

Now when the prey population is at equilibrium $A^* = A_t = A_{t-1}$, then

$$R_{ma} - s_a A^* - vB = 0$$

and, therefore, the equilibrium line for the prey is defined by

$$A^* = \frac{R_{ma}}{s_a} - \frac{vB}{s_a}$$

and we see that the slope of the line, or the marginal cost of predation, is $W_b = v/s_a$. We can also substitute K_a for R_{ma}/s_a to give

$$A^* = K_a - K_a vB/R_{ma}.$$

This equilibrium line will intercept the prey abscissa at K_a when the density of the predator is zero, and the predator ordinate at R_{ma}/v. Thus the predator density, which reduces the prey population to zero, is $P_b = R_{ma}/v$.
The equilibrium line for the predator is defined by

$$1 - B^*/eA = 0$$
$$B^* = eA$$

which is a straight line with slope e. Given a set of parameter values, we can use these equilibrium equations to construct a superimposed reproduction plane and evaluate it graphically. We can also solve the equilibrium system by substituting eA^* for B in the prey equation and $K_a - K_a vB^*/R_{ma}$ for A in the predator equation to obtain

$$A^* = \frac{K_a}{(1 + K_a ve/R_{ma})}$$
$$B^* = \frac{eK_a}{(1 + K_a ve/R_{ma})}$$

Although the stability of this equilibrium is not easily solved with our elementary methods, we can perform numerical steady-state analyses.

This particular model of predation can be written in continuous time as

$$dA/dt = r_{ma}(1 - A/K_a - iB)A$$
$$dB/dt = r_{mb}(1 - B/jA)B$$

in which form it is identical to the model proposed by P. H. Leslie (*Biometrica*, vol. 35, p. 213, 1948). It should be noted that the Leslie model is globally stable over all parameter space, whereas the discrete-time analogue may be unstable if v or e is large due to the time delays inherent in discrete-time systems.

There are a number of other linear predator-prey models (see the disk that comes with this book for their simulations), perhaps the most famous being the so-called "Lotka-Volterra" equations (see Note 4.4 for reference). These can be written

$$dA/dt = (r_{ma} - mB)A$$
$$dB/dt = (nA - d)B$$

As you can see, the death rate of the prey is only affected by the density of its predators - there is no self-limitation - and the birth rate of the predators is proportional to the prey's density alone rather than to the ratio of predators to their prey as in Leslie's model. The "Lotka-Volterra" equations can also be evaluated graphically and, if we do this, we will find that they produce limit cycles with amplitude determined by the starting conditions. The model can be made more reasonable by including a self-limiting expression in the prey equation; damped-stable solutions are then possible.

A more general model has been proposed by M. L. Rosenzweig and R. H. MacArthur (*American Naturalist*, vol. 97, p. 209, 1963), which includes the predator's functional response in the prey equation. However, an analysis of these different formulations of the problem will not shed any new light on the predator-prey interaction. For those interested in further discussion of simple predator-prey models, they are covered in J. Maynard Smith's *Models in Ecology* (Cambridge University Press, 1974). Numerous other models of increasing complexity abound in the literature. Some of these have been summarized in M. P. Hassell's book *The Dynamics of Arthropod Predator-Prey Systems* (Princeton University Press, New Jersey, 1978).

4.10. *Predator-Prey Systems* (see Note 4.9 for complete reference).

4.11. M. E. Solomon (*Journal of Animal Ecology*, vol. 18, p. 1, 1949) was apparently the first to coin the term "functional response" to describe the changes in numbers of prey attacked by individual predators as the density of the prey population changes. However, it was C. S. Holling [*Canadian Entomologist*, vol. 91, p. 385, 1959; *Memoirs of the Entomological Society of Canada*, nos. 45 (1965) and 48 (1966); and subsequent contributions] who investigated the functional response and its components in great detail. Holling also identified the three basic types of functional responses (see Figure 4.19 for review).

4.12. We can calculate the proportion of the prey population surviving predation, p (percentage survival is, of course, $100p$), by

$$p = (A - A_\mathrm{a})/A = 1 - A_\mathrm{a}/A,$$

where A is the density of the prey population prior to predation and Aa is the number attacked in a unit time interval. The proportion surviving from the action of predators with different functional responses can be computed as follows:

Type I: Given a constant attack rate then the number of prey attacked by predators with type I responses is

$$A_\mathrm{a} = \alpha B A \qquad \text{when } A < A_\mathrm{s} B$$

and

$$A_\mathrm{a} = \beta B \qquad \text{when } A > A_\mathrm{s} B \quad \text{or} \quad A = A_\mathrm{s} B,$$

where α is the rate of attack per predator, B is the density of predators. A_s is the prey density needed to saturate each predator, and β is the saturation constant. Allowing the predator density to be unity and substituting in the first equation we get

$$p = 1 - \alpha A / A = 1 - \alpha \qquad \text{when } A < A_\mathrm{s}$$
$$p = 1 - \beta / A \qquad\qquad \text{when } A \geq A_\mathrm{s} B$$

which means that survival is constant $(1 - \alpha)$ when $A < A_\mathrm{s}$ and that it increases with prey density above the threshold A_s (Figure 4.19B).

Type II: If a predator requires t time to handle each prey after it is attacked then, of the total time T exposed to the prey, it spends

$$T - tA_\mathrm{a}$$

time in actual search and capture activity. The proportion of the time that is available for search and capture thus becomes

$$(T - tA_\mathrm{a})/T = 1 - \delta A_\mathrm{a},$$

where $\delta = t/T$. We can insert this expression into the linear type I response to account for predators which spend time handling their prey. Thus,

$$A_\mathrm{a} = \alpha(1 - \delta A_\mathrm{a}) B A$$

or, on rearranging

$$A_\mathrm{a} = \frac{\alpha B A}{1 + \gamma B A} \qquad \gamma = \alpha \delta$$

gives us the well-known "disc" equation derived by C. S. Holling (Note 4.11). Substituting in our survival equation, with predator density at unity, we get

$$p = 1 - \frac{\alpha A}{A(1 + \gamma A)} = 1 - \frac{\alpha}{1 + \gamma A},$$

which shows that the proportion of the prey surviving increases with prey density (Figure 4.19B).

Type III: Leslie Real, in an article in the *American Naturalist* (vol. 111, p. 289, 1977), suggested a generalized functional response model based on the analogy between predator-prey interactions and enzyme kinetics. In this sense enzymes can be thought of as "predators" on their substrates. Carrying the analogy a step further, we can differentiate between allosteric enzymes, which become more efficient at utilizing their substrate as its concentration, or "density," increases; and nonallosteric enzymes, which have a constant efficiency. The kinetics of nonallosteric enzyme reactions can be described by the well-known Michaelis-Menten equation

$$A_a = \frac{\varepsilon A}{A_i + A},$$

where A_a is the amount of substrate utilized, or the number of prey attacked, ε is the maximum efficiency of the enzyme, or the maximum attack rate of the predator, A is the substrate concentration, or prey density, and A_i is an affinity constant which specifies the substrate concentration, or prey density, at which the enzyme efficiency, or attack rate, is half of its maximum; i.e., $A_a = \varepsilon/2$. This equation is, in fact, identical to Holling's "disc" equation with $\varepsilon = \alpha/\gamma = 1/\delta$ and $A_i = 1/\gamma = 1/\alpha\delta$. However, the enzyme kinetic equation is easily generalized to allosteric reactions which give rise to type III responses. In this form

$$A_a = \frac{\varepsilon A^n}{A_i^n + A^n}$$

where n is interpreted as the number of encounters between the predator and prey required before the predator reaches its maximum efficiency (in enzyme kinetics it is the number of binding sites on the enzyme molecule). The parameter n can be thought of as representing the learning ability of the predator, or the rate at which it acclimatizes to a particular prey species out of its repertoire. We can see that, when $n = 1$, this equation reduces to the Michaelis-Menten equation, or the Holling "disc" equation.

We can now insert this generalized model into the prey survival equation to yield

$$p = 1 - \frac{\varepsilon A^n}{A(A_i^n + A^n)} = 1 - \frac{\varepsilon A^{n-1}}{A_i^n + A^n}.$$

From this equation we can see that when $n > 1$ then the survival of the prey will decrease with prey density until $A = A_i$, but that after this threshold prey survival will increase with density (Figure 4.19B). The student is encouraged to demonstrate this numerically.

4.13. Holling originally proposed the type III functional response for animals with learning abilities, particularly birds and mammals (see Note 4.11 for references). These general predators learn that a particular prey species is available and palatable when they encounter it fairly frequently. They then tend to search for that species in preference to others and their rate of attack on it increases; that is, they switch to the more abundant species in their prey repertoire. However, as Hassell points out in the *Journal of Animal Ecology* (vol. 35, p. 65, 1966), specific invertebrate predators with type II responses, but which can only attack prey after they reach a certain density, may produce comparable effects. In addition, S-shaped responses have been observed in other invertebrate predators (e.g., D. G. Embree in the *Canadian Entomologist*, vol. 98, p. 1159, 1966), which suggests that they may be more common in nature than meets the casual eye.

4.14. We can define the reproduction plane for a prey species under predation by an animal with an S-shaped functional response by inserting the type III response (Note 4.12) into our elementary population model [equation (2.7)] to give

$$A_t = A_{t-1} + R_m \left(1 - \frac{A_{t-1}}{K}\right) A_{t-1} - \frac{\varepsilon B A_{t-1}^n}{A_i^n + A_{t-1}^n} .$$

In this model the predators feed on those prey that survive the other density-dependent factors. It is worth noting that v, the vulnerability of the prey to attack of our original model, can be defined as

$$v = \frac{\varepsilon A^{n-1}}{A_i^n + A^n}$$

and we see that vulnerability increases as prey density rises towards A_i, but decreases thereafter.

At equilibrium, where $A_t = A_{t-1} = A^*$, this equation reduces to

$$0 = R_m \left(1 - \frac{A^*}{K}\right) - \frac{\varepsilon B A^{*n-1}}{A_i^n + A^{*n}}$$

The equation can be solved for A^* (see the paper by D. Ludwig, D. D. Jones, and C. S. Holling in the *Journal of Animal Ecology* vol. 47, p. 315, 1978, in which a model for the spruce budworm is analyzed). However, it is much easier to solve the equation for B given a particular A^*; that is, we can ask the question "What predator density must be present in the system given that we have a known equilibrium prey density?" Rearranging our previous equation we have

$$\frac{\varepsilon BA^{*n-1}}{A_i^n + A^{*n}} = R_m\left(1 - \frac{A^*}{K}\right)$$

$$B = \frac{R_m}{\varepsilon A^{*n-1}}\left(1 - \frac{A^*}{K}\right)(A_i^n + A^{*n})$$

and we can see that when $A^* = K$, or the prey density is at carrying capacity, then the predator density must be zero because the second term of the equation is zero. However, when $A^* < K$, then $B > 0$ because all the terms of the equation are positive. If we calculate a set of predator densities necessary to maintain – a series of different prey equilibria, then we can draw a reproduction plane similar to that shown in Figure 4.20B. The only conditions for the system to have two potentially stable equilibria are that $n > 1$ and $K > 5.196A_i$. The serious student is encouraged to compute at least one such reproduction plane.

4.15. An interesting demonstration of the effect of the prey's saturation density, K_a, on the stability of a predator-prey system can be found in J. Maynard Smith's book *Models in Ecology*, published by Cambridge University Press, 1974. On pages 33 to 35 he discusses experiments performed by L. S. Luckinbill with *Paramecium* (prey) and *Didinium* (predator). Luckinbill was able to stabilize an otherwise unstable interaction by cutting the prey's food supply in half. This operation reduced K_a to less than one-half of its previous level and probably had little effect on P_b.

For those interested in a more analytical approach to the problem of predator-prey interactions, which – nevertheless, arrives at much the same conclusions as we do, Maynard Smith's book is recommended.

4.16. There are many examples of unstable predator-prey interactions in simplified laboratory environments. Perhaps the best known are G. F. Gause's early experiments with *Paramecium* and its predator *Didinium* (see Note 4.4 for reference), and C. B. Huffaker's beautiful series of experiments with predator and prey mites (*Hilgardia*, vol. 27, 1958). Huffaker was also able to show that hampering the predators or facilitating the prey's escape tended to stabilize the interaction (see Figure 4.22).

Unstable predator-prey interactions are also seen in nature, particularly in highly simplified agro-ecosystems. An example of an unstable interaction between mite predators and their prey in Washington apple orchards is documented by S. C. Hoyt (*Journal of Economic Entomology*, vol. 62, p. 74, 1969). However, this interaction was more stable when alternative food, in the form of different species of mites, was available for the predators.

4.17 Aphidophagous predators, like ladybirds, syrphids and chrysopids commonly occur in agricultural crops, wild herbaceous plants, and other habitats. Many species feed on aphids and their efficiency in controlling them is a broadly discussed and controversial issue. Their prey, aphids, live in colonies, which are characterized by an initial rapid increase followed by an equally rapid decline in abundance resulting in extinction of the colony. The decline is not caused by aphid predators or parasites, even if they contribute to it. Instead, aphids cause the decline by themselves: they strongly react to their own density by switching to production of migrants that look for another, more suitable host. Thus, when the aphid density is high, most of the newborns leave the colony immediately after reaching adulthood.

The dynamics of different colonies is not synchronized in time, as they feed on different host plants with different phenologies. On a large spatial scale, at any instant, populations of aphids exist as patches of prey, associated with patches of good host plant quality. That is, aphid predators exploit patches of prey that vary greatly in quality both spatially and temporally and therefore have to find a strategy, how to optimally exploit them.

The adult aphid predator is winged, can easily move between patches, and therefore has no problem in finding patches with enough prey items. Thus energy is not a limiting factor for its fitness. Its immature stages are confined to one patch and if this contains few prey items, the larvae starve and cannibalize: eat each other. Mortality of immature stages due to starvation, cannibalism or intraguild predation is enormous: 98–99% and is mainly a consequence of low prey numbers at any time during the larval development. Thus egg and larval cannibalism is adaptive in these predators, as eating conspecific competitors will increase the likelihood of survival of a predatory larva.

Because of the immense egg and larval mortality, selection acts mainly on optimum oviposition strategies – those that insure the maximum likelihood of survival of the offspring – rather than on maximization of the food eaten by the predator per unit time, as considered in most optimum foraging theories (see the book Foraging Theory by D. W. Stephens, and J. R. Krebs, published by Princeton Univ. Press, Princeton, NJ in 1986 for a comprehensive explanation of optimum foraging theories). The optimum oviposition strategy of the adult is therefore determined mainly by expectations of future bottlenecks in prey abundance, as these will affect survival of its offspring, and not by the present amount of prey in the patch, as the adult is not limited by the amount of food – it can find another colony, if needed.

A good long-term forecast of the quality of the prey colony becomes especially important for the ovipositing predator, if the ratio of generation time of the predator to that of the prey (generation time ratio, GTR) is large. This is because the oviposition strategy of a predator with a long larval developmental time will depend on a longer projection of the future prey abundance in the patch, will therefore include more bottlenecks or higher probability of a bottleneck than that of a predator with a short developmental time, and consequently must be more conservative in terms of preserving their prey. As a result, such predators tend to be less effective in controlling their prey. This "GTR hypothesis" seems to hold more generally and those interested in more details are referred to the papers by P. Kindlmann and A. F. G. Dixon: When and why top-down regulation fails in arthropod predator-prey systems (Basic & Appl. Ecol., vol. 2, pp. 333–340, 2001) and Generation time ratios - determinants of prey abundance in insect predator-prey interactions (Biol. Control, vol. 16, pp. 133–138, 1999) and references therein.

Aphid predators are a good example of this hypothesis, as their developmental time often spans several aphid generations, during which the aphid numbers vary dramatically. Laying eggs in the presence of conspecific larvae is strongly selected against in these predators, because it results in these eggs being eaten by older conspecific larvae. In addition, laying eggs late in the development of the ephemeral patch of prey is maladaptive, as there is insufficient time for all the larvae to complete their development. Thus, eggs laid by predators late in the existence of a patch of prey are at a disadvantage, as they are highly likely to be eaten by larvae of predators that hatch from the first eggs to be laid.

Empirical data indicate that several different species of aphid predators have evolved mechanisms that enable them to oviposit preferentially in patches of prey that are in an early stage of development and avoid those that are already being attacked by larvae. Females of these species strongly react to the smell of larval tracks of their own species or of other aphid predators by immediate ceasing oviposition and flying away from the aphid colony. This response strongly reduces the number of eggs laid per patch and consequently their effectiveness in regulating the numbers of their prey – aphids. Thus their optimum oviposition strategy, which maximizes the fitness of the *individual*, results in conserving their prey (or low impact on its numbers), exactly, as stated in the main text. Note that for evolution of this strategy no group selection (see the following note) is needed.

4.18. The idea that natural selection may act on groups of organisms, or populations, as well as on individuals is another subject of controversy amongst biologists. The principal proponent of group selection is V. C. Wynne-Edwards, and those interested in this fascinating subject should consult his works (e.g., *Nature*, vol. 200, p. 623, 1963). He has also summarized his ideas in the book *Natural Regulation of Animal Populations* (edited by I. A. McLaren, Atherton Press, New York, 1971). Also in this book is an article by D. Pimentel on the co-evolution of predator-prey systems. Pimentel argues, and presents data to support his arguments, that predator-prey systems evolve stable interactions via genetic feedback, which produces adjustments in the efficiency of the predator, the vulnerability of the prey, and/or their reproductive potentials.

Chapter 5
Interactions in Space

5.1. Introduction

In our original definition of a population (Chapter 2) we set the spatial limits of the system in a rather arbitrary manner. This was done for practical reasons and, in particular, because it is often difficult or impossible to identify the real geographic boundaries of a population. Our population models (Figures 2.13 and 3.9) dealt with the problem of organisms moving into and out of the system by incorporating net migration (immigration-emigration) into the density-dependent process of population regulation. That is, we assumed that net migration changed in response to the density of the population. If we continue with this line of reasoning, we can draw a diagram for the interaction between two populations of the same species that occupy two spatially distinct environments as shown in Figure 5.1. In this model individuals displaced by competitive interactions from population A enter population B through the porous boundary we have set up between them – in reality this boundary is nonexistent. This kind of reasoning is convenient because it allows us to utilize the theories that we developed in the previous chapters to evaluate the dynamics of populations over broad geographic regions. However, as we shall see later, there may be better ways to define the spatial boundaries of population systems.

If we examine the feedback loop with the shortest possible path created by the spatial interaction (the thick line of Figure 5.1), we find it is overall positive: $E_a \xrightarrow{+} N_b \xrightarrow{+} E_b \xrightarrow{+} N_a \xrightarrow{+} E_a$. For instance, an increase in emigrants from population A increases the density of B, which causes increased immigration into A, and so on, which means that, if migration is the only density-dependent mechanism, then both populations will grow continuously. This observation illuminates the intuitively obvious fact that, when we consider the population over its entire range of distribution, the only processes governing its numerical dynamics are the birth and death rates. Migration only acts to spread the population over its range and to moderate the density of the population in any particular part of this range.

We can also see from Figure 5.1 that immigration from adjacent populations may disturb the balance between a population and its environment. Even if a

A.A. Berryman, P. Kindlmann, *Population Systems: A General Introduction*
© Springer Science+Business Media B.V. 2008

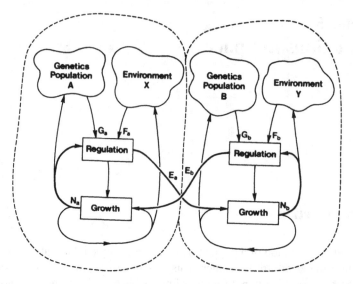

Fig. 5.1 A general model for two populations of the same species interacting with each other across arbitrary spatial boundaries through the movement of individuals. Symbols subscripted by population: genetic properties, G; environmental favorability, F; population density, N; and emigrants, E

population is in a steady state at its saturation density, a large immigration may seriously denude the environment and start off a series of oscillations or cycles. Later in this chapter we will see that this disturbing influence of immigration is extremely important to the dynamics of population systems and, in particular, to the spread of pest epidemics.

5.2. Movements in Space

In order to understand how migration influences the spatial dynamics of populations, we need to look more closely at the movement of organisms and the properties that affect these movements. Most organisms, particularly animals, have some locomotory ability. The most obvious, of course, are those with walking and flying appendages, but many less endowed creatures use wind, water, or other more mobile organisms to disseminate them and their spores. From an ecological viewpoint, we are more interested in what causes organisms to move rather than in how they do it. Organisms with locomotory ability may move for numerous reasons, but they can be conveniently grouped into (1) those movements that are stimulated by the interaction, or the need for interaction, with their own kind, and (2) those that are stimulated by interaction with the environment.

Considering the first group, we may find individuals moving towards each other (+) for the purposes of mating, attacking their prey, or defending themselves. Such movements will usually result in *aggregations*, or clumpings, of individuals.

Conversely, individuals may move away from each other (−) because of antagonistic confrontations with their fellows over mates, territories, and the like, and this will result in the *dispersal* of individuals into a more uniform spatial distribution. In other cases individuals may be completely indifferent (0) towards each other, in which case their distribution will be random in respect to their fellows.

Of course organisms also move in response to conditions they encounter in their environments. In general we would expect them to move away from unfavorable environments and toward favorable ones. Some animals have very complex behavioral mechanisms, which they use to locate favorable conditions for survival and reproduction. Even if direct sensory mechanisms are not involved, organisms will gravitate toward favorable conditions. For example, movements to escape predators, to avoid competitors, or to find food and nesting places will take individuals away from unfavorable localities, whereas the lack of movement in more benign environments will tend to keep them there.

Environments are rarely constant in time and many undergo severe seasonal changes. Migration is one mechanism that enables animals to avoid the unfavorable seasons. In some cases migration may take the animal to distant places, and the navigational problems inherent in this kind of strategy have produced complex orientation behaviors in some migratory animals. Other less mobile species are forced to remain in place during the unfavorable periods, but avoid the problem by hibernating when the weather becomes too cold or aestivating when it becomes too hot.

As we know, organisms may also affect the favorability of their environments by overexploiting needed resources or by pollution. In such cases they will tend to move on as the environment deteriorates. Aphids are a typical example of this type of behavior (see Note 5.13).

The movements of organisms in response to each other and to their environments causes them to assume patterns in space that are characteristic of that particular species. These patterns will change in time as the environment changes or as the response between individuals changes. Some examples of the type of patterns we encounter in nature are given in the following paragraphs.

1. Organisms that respond positively toward each other usually form aggregations into herds, flocks, schools, or swarms. As is often the case with such aggregations, the environment may be severely overexploited and the herds continuously on the move so that the species lives a *nomadic* existence (Figure 5.2A). Many grazing ungulates, some of our flock-tending ancestors, and insects like the locust, exemplify this way of life.

2. Antagonistic interactions are usual in *territorial* species. In this case an individual or pair marks out a territory, which is more or less fixed in space, and defends it against challenges by others of the same species (Figure 5.2B). Territorial animals usually move only within the borders of their own territory or home range and, therefore, they tend to be rather uniformly distributed in space. However, as territories will usually be larger in less favorable environments, the density of the population will be affected by conditions in the environment.

3. We frequently observe mixtures of patterns (1) and (2). For example, some species form aggregations that are antagonistic toward other colonies of the same

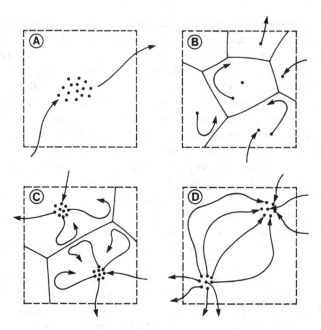

Fig. 5.2 Some patterns and movements of populations in arbitrarily delimited space: (A) Aggregated population moving in unison in response to environmental conditions, (B) territorial pairs. (C) dispersed aggregations, and (D) dispersal at certain times and aggregation at others in response to environmental and individual variations

species – ant nests, lion prides, monkey colonies – which will result in dispersed aggregations as shown in Figure 5.2C. Other species form nomadic aggregations at certain times and others disperse into territories (e.g., some territorial birds and mammals).

4. Some organisms are rather indifferent toward each other, but even so will rarely be distributed randomly in the environment. Clusters will be formed by breeding pairs and patches denuded by predators to create a nonrandom but haphazard mosaic in space. However, movement in response to environmental conditions will then tend to even out the distribution again, because dense groups will overexploit and sparse groups underexploit the environment. Thus, we will see a continuously shifting pattern as the environment changes in response to the exploitation of the population and the movement of its predators and competitors.

5. Mixtures of patterns (1) and (4) will also be observed, particularly in *opportunistic* species occupying temporary habitats (the bark beetles discussed in Note 3.5 and aphids are a good example). Such species, which are indifferent toward each other at certain times but which aggregate at others, create complex spatial patterns in which aggregations in one place disappear to reappear in a different place (Figure 5.2D). Other species, such as whales, seals, and salmon, aggregate at certain times for mating purposes, but then disperse over much larger feeding areas.

Exodus from a particular environment usually occurs when needed resources are depleted, or when the environment becomes intolerable because of physical conditions or the presence of other organisms (competitors, predators, or pathogens). Thus, dispersal is extremely important for those species that denude their resources or that inhabit very variable environments. It is not surprising, therefore, that animals with highly developed dispersal powers, such as the birds and insects, have been most successful in utilizing rare or temporary habitats. Birds, for example, with their powerful flight and navigational abilities, have exploited tundra and arctic environments, which are only favorable for short periods of time each year. Insects have evolved equally intricate physiological and behavioral traits for dealing with the problems of scarce resources and changing environments. For example, bark beetles use mixtures of volatile chemicals to guide their brethren to individual weakened trees; female aphids reproduce wingless offspring in favorable environments, but winged forms are produced when overcrowding depletes the food supply (see also Note 5.13); and locusts go through remarkable changes in physiology and form under crowded conditions, which lead to mass exodus from the overexploited environment and the terrible migratory swarms which devastate all in their path.

In contrast, organisms inhabiting consistently favorable environments tend to be less mobile. In fact, dispersal may be disadvantageous for such species because their energies are better spent in reproduction, care of their young, outcompeting their rivals, or in defenses against predators and pathogens. This point is made clear by animals that have evolved poorer dispersal powers than their ancestors; for example, the ostrich living in the uniform and endless Sahara and the wingless Douglas-fir tussock moth inhabiting the extensive fir forests of western North America. However, organisms without good dispersal powers cannot afford to overexploit their environments and must, therefore, practice conservation by limiting their own numbers. Territorial behavior is one of the most successful tactics for achieving these ends.

Southwood (Note 5.1) has argued that the evolution of life-cycle strategies is closely tied to the spatial and temporal variability of the environment in which an organism lives. As we have seen, dispersal and migration are tactics employed by organisms occupying spatially variable environments, while temporal variations can be dealt with by entering a dormant stage which is resistant to the unfavorable conditions: hibernating bears, diapausing insects, and fungal spores. Thus, the life-cycle strategies of some organisms (particularly insects subjected to temporal and spatial variations in their environments) may be extremely complicated. As Southwood emphasizes, the environment, or habitat, acts as a template within which the life-cycle strategy of an organism evolves so that breeding, dormancy, and dispersal occur when they are most beneficial for the reproduction and survival of the species. These ideas are summarized in Table 5.1.

Table 5.1 Life-Cycle Tactics for Dealing with Environmental Variations in Time and Space[a]

		Time environment favorable	
		Now	Later
Place environment favorable	Here	Breed	Enter dormancy and breed later
	Elsewhere	Disperse and then breed	Disperse, enter dormancy, and then breed

[a] Modified from Southwood (Note 5.1).

5.3. Dynamics in Space

We can construct a model of populations interacting over extensive geographic areas by dividing the area into a large number of compartments, or a grid, and then assume that the rules of population growth and regulation apply within each compartment. In other words, the subpopulations inhabiting each grid element are assumed to operate independently of each other except for the movement of individuals between them. It is also necessary to assume that the environment within each compartment is homogeneous and can be classified according to its favorability for reproduction and survival of the organism. Although this approach has some undesirable features, which we will try to correct later in this chapter, it is the most commonly used method for evaluating the dynamics of populations in space.

If we set up a spatial grid, and given a starting distribution of individuals over this grid, then the dynamics are evaluated as follows: First, the subpopulations within each compartment grow according to the rules derived in Chapter 3 (Figure 3.9). However, the population regulation process must be subdivided into a density-dependent birth and death process, and a density-dependent *emigration* process governing the number of organisms leaving the compartment. Because the condition of the environment determines the intensity of competition, it also controls the rate of movement out of the compartment. Hence, the movement of individuals in response to their environment is an intrinsic part of the model.

The direction of movement out of compartments will be very important in determining which adjacent compartments receive immigrants. In the simplest case emigrants may leave in equal numbers across all four boundaries. However, directed movements across one or two boundaries may occur if the population forms into nomadic aggregations or if weather conditions influence the direction of movement - for example, small flying insects are affected by wind, rain, and temperature.

The immigration of individuals into a particular compartment will depend on its distance, and perhaps its bearing, from grid elements that are sending out migrants.

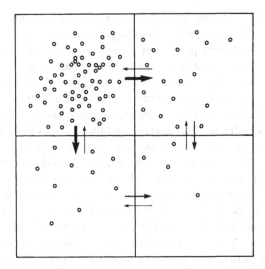

Fig. 5.3 The direction of density-dependent migrations between subpopulations occupying adjacent compartments in space; the thickness of the arrows indicates the magnitude of the movements across the boundaries

Of course, organisms with well-developed locomotory abilities will travel further than more sluggish ones, and conditions in the environment such as atmospheric and ocean currents may also be important. However, the net result of the emigration process will usually be a movement of organisms from the crowded compartments and into the less crowded ones in the more favorable environments (Figure 5.3).

Once the immigrants have entered a compartment they may pass through the population growth process if they are reproductively mature, but they will have the characteristic individual rate of increase of their home compartment. Immature individuals will bypass this and enter the population regulation process, which may cause them to die, become mature and pass on to the growth box, or emigrate again.

As you can see, spatially defined population models can become extremely complicated and, because of this, analytical solutions are often difficult or impossible to obtain (however, see Note 5.2). Instead we usually have to resort to computer simulations in our studies of populations inhabiting broad geographic regions. Perhaps the most elementary example of a spatially defined system is the "Game of Life" we discussed in Chapter 1, which you can access and run using the disk that comes with this book. Here we saw that the spatial arrangement of individuals was very important in determining whether the "population" grew, remained static, or declined to extinction (Figures 1.15 and 1.16). There have also been some interesting simulation studies with more realistic population models. We will examine one such study below.

The spruce budworm is a moth that feeds on the foliage of spruce-fir stands over much of North America. The insect population is normally kept at very low densities by the action of its predators and the lack of favorable environments – a favorable environment for the budworm is a dense forest of mature balsam fir, at least in the eastern part of its range. Once every 30 to 50 years, after a supply of mature firs has accumulated, the budworm population escapes from its predators, explodes to extremely high densities, and kills most of the mature firs in the forest. There is evidence that the main budworm predators have S-shaped functional responses which create an N-shaped prey equilibrium line and a predator-prey interaction similar to that of Figure 4.21A. The budworm is able to increase above the unstable threshold in years with warm, dry springs, and then escapes from the regulation of its predators. Thus, the population eruption seems to be initiated by the coincidence of mature balsam fir stands and favorable weather conditions. Once the outbreak has been triggered the budworm rapidly devours the fir's foliage and kills most of the mature trees. The population then collapses back to a very low density.

Researchers at the Institute of Resource Ecology (University of British Columbia), the Canadian Department of the Environment, and the International Institute of Applied Systems Analysis have assembled the mass of information on the budworm interaction with spruce-fir forests into a spatially defined model of the Province of New Brunswick (see Note 5.3). The Province was divided into 265 compartments of approximately 66 square miles and each compartment was classified according to its favorability for the budworm. The dynamics within each compartment were governed by interactions between the budworm, its predators, and forest and weather conditions as we briefly outlined above. Emigration and immigration rates were controlled by density-dependent processes and the flight characteristics of the adult moth. Simulations on this model produced pictures of the space-time dynamics such as that shown in Figure 5.4. Outbreaks occurred with a periodicity similar to the natural outbreak cycle, and the model proved useful for evaluating the effectiveness of various management alternatives.

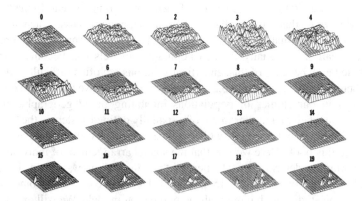

Fig. 5.4 Simulation of spruce budworm population dynamics in New Brunswick over a 19-year period (reproduced by courtesy of the authors from a chapter by W. C. Clark, D. D. Jones, and C. S. Holling in the book *Spatial Pattern in Plankton Communities*, edited by J. H. Steele and published by Plenum Press, New York, 1978)

There have been other simulation studies of population dynamics in space, but the spruce budworm example will suffice to illustrate the grid approach. We will, however, draw on these other examples when we discuss the ecological implications of spatial interactions in the following paragraphs.

5.4. The Spread and Collapse of Pest Epidemics

The spruce budworm is a particular case of a much more general problem, which has plagued man from time immemorial – the spread and collapse of pest epidemics. Pests are organisms that have a negative effect on man's survival and well-being, either as competitors for resources (like the budworm) or as predators and parasites of man. Although man, with his technological prowess, outwitted his larger predators a long time ago, he has had much more difficulty dealing with many disease-causing microorganisms and with the competitors that ravage his supplies of food and fiber.

Many pest organisms remain for long time periods at low, or *endemic*, population levels, and are often tolerated by man when in this condition. However, at certain times and under certain conditions the pest populations erupt into *outbreaks* or *epidemics*, which may cause widespread discomfort (flu virus), loss of life (plague bacteria), or destruction of food (aphids) and fiber (spruce budworm).

Because of the general nature of the problem of pest epidemics, it is worth looking a little more deeply at the spruce budworm system. First, consider the interaction between the budworm and the forest in the absence of bird predators. The condition of the forest, particularly its species composition, density, and maturity, determines the favorability of the budworm's environment and we might expect a reproduction plane such as that in Figure 5.5A (the exact form of this plane is unimportant to the general problem we are investigating). As the forest grows and matures, the environment becomes more favorable for the budworm and its population will increase to higher equilibrium densities, but because forest growth is a very gradual process, we would not expect any dramatic population changes. Now in the presence of a relatively constant population of bird predators the equilibrium budworm population will be suppressed, at least until the functional responses of the birds begin to saturate. When this occurs, bird predation will no longer act as a negative feedback mechanism (see Figure 4.20 in Chapter 4) and the budworm will escape from its endemic level. In a slowly maturing stand of balsam firs the budworm population will be maintained at an endemic level by bird predation, but as soon as the stand conditions permit sufficient reproduction and survival to disengage the negative predation effect, the population will erupt towards its upper equilibrium position (Figure 5.5B). Because of the explosive growth of this released population it will probably exceed its upper equilibrium level by a large margin, denuding its host trees of their foliage and killing many of them. In effect the budworm causes a drastic reduction in the favorability of its environment and the inevitable collapse of its own population (Figure 5.5B). We can also see how changing

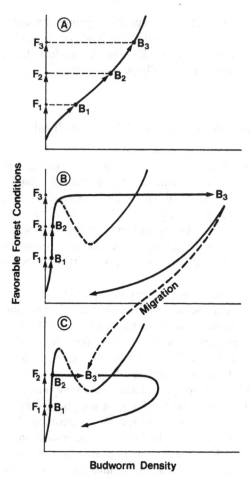

Fig. 5.5 (A) Hypothetical reproduction plane for the spruce budworm in the absence of its avian predators, (B) the plane when a constant number of bird predators are present showing an endemic-epidemic trajectory, and (C) the triggering of an epidemic by immigrations from a nearby area. B_1, B_2, ..., etc., indicate budworm densities at time 1, 2, ..., etc., as forest conditions change, or following immigrations from surrounding forests

weather patterns can act as a trigger for the epidemic. For example, suppose that the population is at B_2 in Figure 5.5B when a period of favorable weather occurs. The population will be carried around the apex and explode. A return to more normal weather conditions, or even to unfavorable ones, will have no appreciable affect on the epidemic trajectory once it has passed the unstable threshold (the broken line in Figure 5.5B).

We can also see how emigrants from overcrowded outbreak areas can spread the epidemic into adjacent regions. For example, we would expect large numbers of emigrants from population B_3 (Figure 5.5B) because the food supply will have been denuded and the environment made very unfavorable. When these emigrants land in other forested areas they will be added to the resident population and may raise its density above the outbreak threshold (Figure 5.5C). In this way an outbreak *epicenter* may become the match, which starts a widespread conflagration. The epicenter concept is extremely important to applied ecologists because it implies that epidemics can be controlled or prevented by treating relatively small areas (the epicenter), which is an alternative to the large-scale and costly treatments that are often required to control rampant epidemics (see also Note 5.4).

There are certain features that are common to all population systems that exhibit eruptive epidemic behavior, no matter whether they are forest insects or human pathogens. The most important is that they all possess critical *thresholds* separating endemic from epidemic behaviors. Of course, understanding the mechanism that determines the threshold is crucial if we are to control pest epidemics. In many cases the threshold is related to the ratio of susceptible to immune individuals in the host population. Using these terms in their broadest sense we can see that a spruce budworm outbreak can be triggered when the ratio of "susceptibles" (mature balsam firs) to "immunes" (spruces, hardwoods, and immature firs) reaches a critical level – in other words, the environment becomes very favorable (Figure 5.5B). In a similar fashion, disease epidemics usually erupt when a large proportion of the population is susceptible to infection. This may occur when a large number of susceptibles migrate into an area, when explosive population growth gives rise to a large number of individuals that have not been previously exposed to the pathogen, or when a few infected individuals migrate into a susceptible population. Migration, as we see, plays just as important role in the epidemiology of diseases as it does in forest insects. This is nowhere better illustrated than in man's migrations, which are strewn with the victims of disease epidemics (see Note 5.5).

The concept of a *critical population threshold* separating low-density endemic dynamics from the devastation of a pest epidemic can be extremely useful to the pest manager. If the threshold can be identified, then it presents the opportunity for predicting pest outbreaks. For example, epidemics of the mountain pine beetle are related to the favorability of the environment for the reproduction and survival of the beetle (particularly the thickness of the inner bark of lodgepole pines where the beetles live) and to the vigor of the trees, which determines their ability to resist the beetle attack (Note 5.6). In this case the vigor of the lodgepole pine stand acts, much like bird predation on the spruce budworm population, to prevent the beetles from utilizing their potential food supply. However, once the beetle population reaches a critical density it is able to overwhelm the defenses of vigorous trees by massive and rapid attack. It is this cooperative activity, which creates the unstable threshold (see Chapter 3). With this information, and data from a number of different lodgepole pine stands, we can find the approximate location of the epidemic threshold as a function of phloem thickness and vigor of the stand (Figure 5.6). Once the threshold function has been derived, the manager can use it to identify those stands in his forests that are most likely to experience a beetle epidemic. He does this by measuring the phloem thickness and vigor of a particular stand and seeing how close it is to the epidemic threshold. Naturally, the nearer a stand is to this threshold

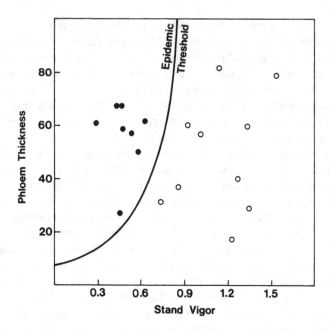

Fig. 5.6 The approximate location of the threshold separating endemic (○) from epidemic (•) behavior of mountain pine beetle populations in lodgepole pine stands of different phloem thicknesses and vigor, where vigor is measured by stand age, density, and periodic growth (see Note 5.6 for reference)

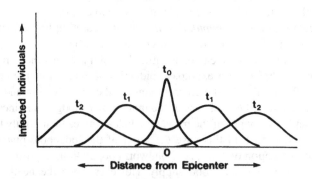

Fig. 5.7 Three states in time of an idealized epidemic wave spreading from an epicenter

the greater is the chance that a minor environmental disturbance, or a beetle immigration, will push it into the epidemic domain.

The second common feature of many endemic-epidemic systems is that the emigration of pests, or infected hosts, from the outbreak epicenter often spreads the epidemic into new areas. This may cause the outbreak to proceed in a wavelike movement through space (Figure 5.7). However, when the organism has highly developed dispersal abilities (e.g., insect pests), when it is transported *en masse* by physical currents (e.g., the spruce budworm on weather fronts), or when the host transports them for long distances (e.g., human pathogens), we may observe new epicenters being formed at some distance from the original outbreak. These new centers may then spread and coalesce into ever-changing patterns of waxing and waning infestations (Figure 5.4).

The third characteristic of many endemic-epidemic systems is that climatic variations from the norm may act as a trigger, which sets off the epidemic wave. This phenomenon is particularly evident when climatic changes affect the susceptibility of the host. For example, flu epidemics often follow an abrupt change in the weather, which places stress on the human host and creates a large population of susceptible individuals. Another example can be found in the bark beetles that attack and kill trees that are under stress. Changes in the normal weather patterns, particularly droughts but in some cases too much rainfall, may lower the vigor of large numbers of trees and the beetle population may explode to epidemic proportions.

The bark beetles illustrate another characteristic of some epidemic systems. As we mentioned previously, when bark beetle populations become large they are able to circumvent the defenses of their hosts. Even quite healthy hosts are unable to deal with a continuous assault by large numbers of beetles. Similar relationships between the *pathogenic load* and the host's ability to defend itself are sometimes found in other systems. For instance, a healthy person may contract a viral infection if he is continuously exposed to infected individuals. In these cases, epidemics that start in susceptible epicenters may spread through more resistant populations.

5.5. Stability in Space

In Chapter 4 we saw that interactions between efficient predators and vulnerable prey may be unstable in any given locality, but this does not mean that the system will be unstable over a large geographic region. In fact we will see that spatial dimensions often act as a stabilizing force in otherwise unstable systems. Once again the spruce budworm model can be used as an example. Take, for instance, a mature stand of balsam firs, which is killed by a budworm epidemic. The insect population will undoubtedly disappear from this locality because there is no food to sustain it. In this particular locality the budworm "predator" is very efficient and the coniferous "prey" highly vulnerable, thus creating an unstable local system. Elsewhere however, budworm populations will be in various stages of the endemic-epidemic cycle and the overall population will persist until the particular stand regenerates and matures to be devastated by subsequent budworm epidemics (Figure 5.4).

The effect of space on the stability of predator-prey systems was also demonstrated by Huffaker's experiments with mite populations (see Figure 4.22, page

132). Huffaker grew populations of herbivorous mites (small spider-like creatures) on individual oranges. When predatory mites were introduced onto or happened to find an orange bearing prey, they quickly exterminated the prey population. Thus, the interaction on a particular orange was unstable because the predator was very efficient and the prey vulnerable. However, when a large number of oranges were dispersed in space and physical barriers were present, which helped the prey to disperse while hindering the predator, the two populations coexisted for long periods of time. Prey populations were still exterminated on oranges that the predators invaded but, in the meantime, immigrants were colonizing uninhabited oranges and insuring the persistence of the system (Figure 5.8). What we observe is a continuous game of hide-and-seek in space, with the prey always one jump ahead of its predators. This result is perhaps intuitively obvious because an empty orange is of no use to the predator, but forms a very favorable environment for the prey. Hence, the prey has the advantage of being the first to colonize a new environment and is able to build up a population before the predator can find it. Of course, if the predator has very strong dispersal and prey-finding mechanisms (i.e., it is very efficient at finding new sources of prey), the time advantage may be very short and the system will be in danger of total extinction. We are saying, in effect, that stability in space is improved when there are long time delays in the predator's response to the spatial distribution of its prey. This is an interesting result because we previously found that time delays create instability at any one place. Now we are countering that argument by suggesting that time delays in space act as a stabilizing force.

The effect of space on the stability of theoretical predator-prey systems has been investigated by J. Maynard Smith. He found that migration had no appreciable affect on stability when both predator and prey moved immediately to adjacent compartments. However, when he simulated a more realistic system, based on Huffaker's experiments, he found that populations that were locally unstable could persist indefinitely in space (Figure 5.9). The chances for coexistence were improved when the grid was composed of a large number of elements and when the prey had good dispersal abilities, which ensured that unoccupied compartments were colonized as rapidly as possible. In effect, when the prey has a high propensity to migrate it increases the jump that it has on the predator and, consequently, the delay in the spatial response of the predator. Maynard Smith's theoretical simulations produce a dynamic behavior in space, which is very similar to the experimental results obtained by Huffaker (cf. Figures 5.8 and 5.9), and both demonstrate the important rule that space and migration have strong stabilizing influences on population systems (see also Note 5.7).

Although most of the experimental and theoretical research on the stability of spatially defined systems has been concerned with predator-prey interactions, movements in space can also act to stabilize competitive interactions (see Note 5.8). The possession of good dispersal abilities can confer advantages to a weak competitor. For instance, the weaker species may be able to colonize new or underexploited environments very rapidly and hold them temporarily through the advantage of

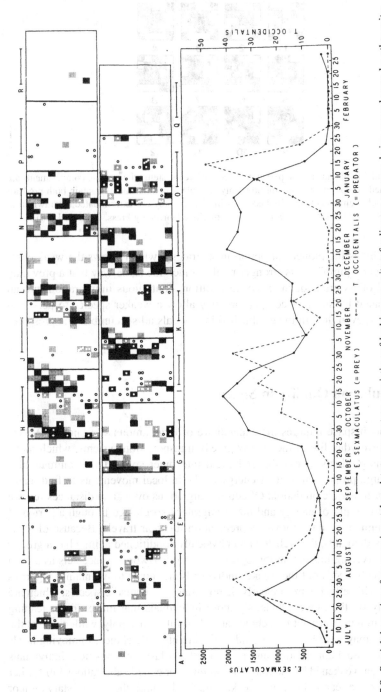

Fig. 5.8 Spatial dynamics of an experimental predator-prey system consisting of herbivorous mites, feeding on oranges, being preyed upon by other mites. The space dynamics are shown in the upper charts, darker shading indicating higher prey densities on individual oranges, and circles the presence of predators. Each space portrait, A through Q, represents a period of time on the lower time scale of length shown by the horizontal lines A to Q (reproduced with the permission of C. B. Huffaker from a paper in *Hilgardia*, vol. 27, p. 343, 1958)

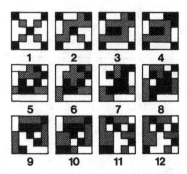

Fig. 5.9 Spatial dynamics of a hypothetical predator-prey system where white compartments are empty, hatched compartments contain prey only, and black compartments contain both predator and prey. Black elements eventually become white as the prey are exterminated (redrawn from J. Maynard Smith's book *Models in Ecology*, Cambridge University Press, London, 1974)

numbers. This is the pioneer or opportunist strategy, which allows the weak competitor to get a "jump" on its stronger rival in much the same way that a prey may keep ahead of its predator in space. In addition, continuous immigration into an area dominated by stronger competitors may allow a weaker species to persist at low numbers, even though the established individuals all succumb to the pressures of competition.

5.6. Population Quality in Space

Travel through space requires the expenditure of large amounts of energy. We see this in our everyday life in the energy-greedy transportation systems, which move man and his products across the globe and into outer space. Other animals also expend a large proportion of their energy intake in local movements to find mates and food, or to escape predation. Of course, migrations over great distances require an enormous output of energy and many migratory species go through a period of energy accumulation and storage in preparation for their travels. Because of these large energy demands, the well-fed and physically conditioned animal has a greater chance of surviving movement through space. In other words, space acts as a kind of sieve, which weeds out the weaker individuals and tends to maintain a vigorous population. This effect is most clearly demonstrated by nomadic species, which are continuously on the move. Obviously those individuals that make up the leading edge of the moving herd get first choice at a plentiful food supply, while those at the trailing edge may have little or no food, and certainly little choice. In moving herds there is a continual competition and jockeying for a favorable position. Individuals that have been weakened by malnutrition, wounds, or genetic aberrations fall further and further to the rear and become weaker and weaker until they die of starvation or are picked off by roving predators, which often trail a moving herd.

Space also acts as a sieve to weed out the weaker specimens in less mobile populations. In these cases individuals tend to emigrate or to produce spores, seeds, and other dispersal forms when the environment in which they live deteriorates. Even if both healthy and weak individuals emigrate in similar numbers, the healthy are more likely to survive to locate and colonize a new favorable environment. However, it is possible that the vigorous, well-fed, and genetically superior specimens will have a greater *propensity* to emigrate. If this is true, and there is evidence to support this view (see Note 5.9), then the new environments will be colonized by these superior individuals while, at the home front, under-nourishment and disease will prevail and the gene pool will deteriorate as the more vigorous genotypes are siphoned off. In addition, the new environment will be relatively free of predators and diseases, at least for a time, while predators and diseases will run rampant in the old environment.

We frequently observe populations of plants and animals, including man, going through cycles of growth, overcrowding, decadence, and sometimes extinction in a particular locality. Our concept of the population operating as a spatially defined system now has the qualities to explain this scenario. Consider an unexploited compartment in space that is occupied by a few hardy pioneers – perhaps those that escaped from an overcrowded and deteriorating compartment some distance away and survived the rigors of migration. Food and space are plentiful, predators and disease scarce, and they flourish and multiply. However, we know these idyllic conditions cannot last for long. The population grows exponentially, and soon competition for food and space lowers the quality of life and the physiological well-being of the individual. The environment deteriorates as the food is used up faster than it can be replaced, predators begin to invade and reproduce, and disease spreads through the crowded population. The weaker members will die in ever increasing numbers while the strong, finding the overcrowded and disease-ridden environment intolerable, will migrate in search of "greener pastures" to become the new generation of pioneers. These pioneers may encounter severe hardships as they search through space for a favorable environment, but this process will ensure that the new colony will be formed by the hardiest individuals.

Meanwhile, at the home front, conditions will go from bad to worse as food is exhausted, pollutants accumulate, and predators and diseases increase (Figure 5.10). Emigration of the vigorous genotypes will leave behind a gene pool that becomes more and more impoverished of "survival" genes. A collapse is imminent, and when it occurs the possibility of local extinction becomes very real. After this final disaster the environment is able to rejuvenate slowly to await the arrival of new pioneers, or to be exploited by the few remaining survivors.

The spatial scene dominated by these migratory patterns will contain young, vigorous, and growing populations; dense populations in which food shortage, disease, and predators are evident; and collapsing populations in severely deteriorated environments. This picture will continuously change in time and space to create a mosaic of waxing and waning populations (e.g., Figure 5.4) with a changing qualitative structure. This scenario, which may be quite common in nature, leaves us with some uneasy feelings when we contemplate the future of mankind. Human

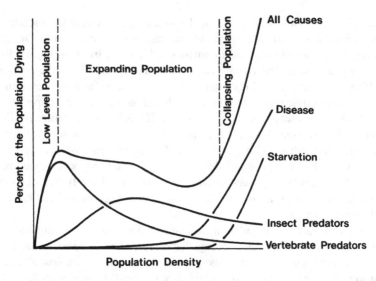

Fig. 5.10 Mortality of gypsy moth larvae caused by various natural mortality agents as the moth population goes from very low to very high densities, showing that mortality from disease and starvation becomes extremely severe in the dense populations. It is also interesting that vertebrate predators (mice, skunks, and birds) cause higher percentage mortality when population density is low, suggesting that their functional responses can create a low-density equilibrium (i.e., an N-shaped prey equilibrium line as shown in Figure 4.20) (reconstructed from a paper by R. W. Campbell entitled "The gypsy moth and its natural enemies," published by the U.S. Department of Agriculture, Information Bulletin No. 381, 1975)

history is filled with examples of the rise, decadence, and collapse of cities, empires, and civilizations, and of the mass migrations of our forefathers to found new and dynamic colonies in new environments (usually, it must be added, at the expense of other civilizations). Now, however, the opportunity to migrate has been all but eliminated by the population explosion of the last two centuries and the delicate balance of power between nations. Without the spatial sieve and migration, what will happen to the quality of our gene pool? Trapped in a more and more crowded environment, what will happen to our physiological and psychological well-being and the quality of our lives? What effects are the accumulation of pollutants and our industrial exploitation having on the quality of our environment? Are some of these effects time delayed so that they will return to haunt our children and our children's children? These questions and many others are not easily answered.

In the preceding sections of this chapter we have seen that population systems that exhibit cyclic instability at any particular place persist in a relatively stable condition over large geographic regions, and that migration plays a critical role in the persistence of the system. However, it should be emphasized that many, if not most, species remain at much more constant densities over long periods of time throughout their ranges of distribution. In these species migration plays a less important role in the space-time dynamics and many have rather sessile habits or, at least, more poorly developed locomotory abilities. These species will usually be generalists or specialists whose energies are channeled into defense against predators and diseases, aggressive territorial behavior, and care of their young, rather than into migration. It is not surprising, therefore, that in our brief look at the operation of population systems in space, our attention has been diverted toward the opportunistic species that is forever reaching into space for new environments to conquer. Man's reach into outer space is perhaps, the ultimate act of an opportunistic species.

5.7. Environmental Stratification

Throughout this chapter we have used the grid approach to evaluate the space-time dynamics of population systems. To do this we had to assume that each grid element contained an environment of uniform favorability for the species in question. However, the boundaries separating environments of differing favorability do not change along grid lines but rather in response to climate and the physical characteristics of the landscape. Climate is the overriding force that molds the environments of all organisms. Naturally, if climate is always unfavorable in a particular area, a species cannot persist there, and so climatic conditions will determine the outer boundaries of the species' distribution in space. Within this broad distributional range, local climate will vary according to topological characteristics of the landscape. For example, the altitude and direction of the slope of the land may cause severe variations in local climates – slopes facing the sun and lower elevations being warmer and drier than their counterparts. These regional climates, together with the local soil and substrate conditions, set the framework within which plant communities evolve, which in turn set the stage for evolving animal communities. Thus, the favorable environment for a predator is much more restricted than that of the herbivore on which it feeds, and the herbivore's is more restricted than its food plant's. This is because they are dependent on the presence of their food resources as well as favorable climatic conditions.

A geographic region can be stratified into zones of favorability for a particular organism if we know how climate, and the plants and animals present, affect the reproduction and survival of that species. This idea has been most fruitful in the classification of plant habitats, a habitat being the environment in which a species normally lives. In this way geographic areas can be divided into zones that favor the growth and reproduction of certain plant associations to produce what are often called habitat-type maps. Plant habitats are recognized by a distinctive combination of vegetation growing on a particular site, and each has a characteristic pattern of development toward a climax community. For this reason, habitat-type classifications have proven useful for predicting the succession of plants that will occupy a given area and the eventual climax association (see Note 5.10). Forest and range managers have been most active in the application of the habitat-type concept.

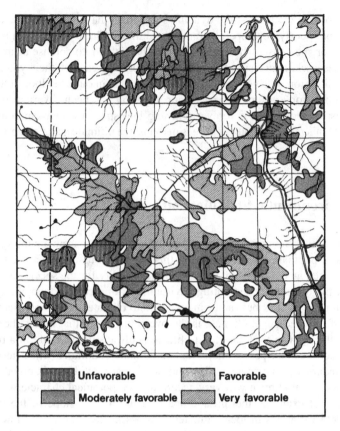

Fig. 5.11 Stratification of part of the Gallatin National Forest into zones of favorability for the mountain pine beetle. The white background represents very unfavorable habitats – forest or other land occupied by non-host plant species. Unfavorable – elevation over 2500 meters or small pines less than 5 inches diameter at breast height; moderately favorable – elevation under 2500 meters, trees 5 to 11 inches DBH and more than 80 years old; favorable – elevation under 2500 meters, trees greater than 7 inches DBH and 80 years old but mixed with other species; very favorable – elevation under 2500 meters, trees greater than 11 inches DBH and 80 years old (redrawn from U.S. Forest Service, State and Private Forestry, Missoula, Montana, Survey Report 76-5, prepared by M. D. McGregor, D. R. Hammel, and R. C. Lood)

However, the same principles are equally useful for classifying the environments of animals.

Some of the more successful attempts at classifying, or stratifying, environments have been done with forest insect pests under the guise of "risk classifications," where a high-risk environment is very favorable for the pest and is, therefore, in danger of being damaged (see Note 5.11). For example, a dense stand of mature balsam fir would be in danger of a spruce budworm outbreak and would, therefore, be classified as high risk. Another example of environmental stratification is illustrated in Figure 5.11. Here an area of the Gallatin National Forest in Montana has been subdivided into zones of favorability for the mountain pine beetle. These beetles reproduce most successfully in the older larger diameter lodgepole pines growing in the lower elevations (see also Note 5.12). These stands are in the greatest danger or, from the point of view of the forest manager, pose the greatest risk of being killed by the bark beetle. As we know from our previous discussion, outbreaks originating in these stands may then spread into the less favorable areas to create a general conflagration.

The stratification of an area according to its favorability for a particular species allows one to see, at a glance, the potential spatial distribution and abundance of that species. Within each zone of favorability the population dynamics should follow a particular trajectory because the environment is roughly homogeneous. The most serious disadvantage of the environmental stratification approach is that rather complex spatial mosaics are produced which are difficult to handle mathematically. However, recent applications of certain mathematical techniques are helping to solve this problem (Figure 5.12).

There are a number of advantages of environmental stratification over the more usual grid approach. First, we can see from Figure 5.11 that the environment is rarely homogeneous within a grid element, and this violates one of the assumptions of the grid method. Second, the stratification method exactly specifies the boundaries and the area of each environmental patch. The size of a particular environmental patch of given favorability is quite important to dispersing organisms because their chances of locating the patch improve as the patch gets larger. In addition, epidemiological studies have shown that patch size is important in the initiation of epidemics; that is, there is a critical patch size below which an epidemic cannot be triggered internally, although it can, of course, be set off by immigration from surrounding areas (see Note 5.2). Lastly, the stratification of forest, range, and agricul-

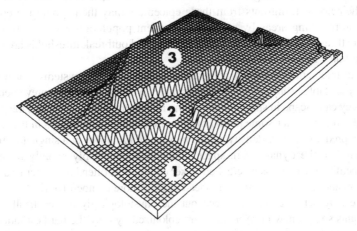

Fig. 5.12 The mathematical representation of an area of forest land according to its favorability for the larch casebearer, a small moth feeding on larch foliage: 1 = low, 2 = moderate, 3 = high favorability. The height of the landscape is a relative measure of the potential carrying capacity for the insect (reproduced with permission from a paper by G. E. Long in *Ecological Modelling*, vol. 8, p. 333, 1980)

tural lands into zones of favorability for particular plant and animal associations provides a basic framework for the management of these populations.

5.8. Chapter Summary

In this chapter we looked at the interaction between populations of the same species inhabiting different environments separated in space. These interactions involve the movement of organisms in space and their immigration into favorable environments and emigration out of unfavorable ones. The major points are summarized below:

1. Individuals move into or out of particular environments as a result of interactions with their own kind or with their environment. These interactions cause the population to assume a particular spatial distribution and pattern of movement through space, which is usually characteristic of the species.
2. Organisms have evolved complex life-cycle strategies for avoiding harsh environments and exploiting variable and temporarily favorable environments. These strategies usually involve dispersal, dormancy, or a combination of both. Species with well-developed dispersal powers have been most successful in exploiting rare or temporarily favorable habitats.
3. Population dynamics in space can be evaluated using a spatially defined grid and a set of rules governing the movement of individuals between grid elements; for example, density-dependent emigration rates, species-specific dispersal characteristics, and environmental properties such as air and ocean currents and measures of favorability.
4. Pest epidemics are usually set off in a particular locality, or outbreak epicenter, by environmental changes that favor the rate of increase of the pest or put stress on their hosts. Emigrants from these epicenters may then spread the epidemic into surrounding areas by raising the resident population above a critical epidemic threshold. The concepts of epicenters and outbreak thresholds are of great importance in pest management.
5. Spatial interactions between populations tend to stabilize systems that might be highly unstable in any one place. For example, in unstable predator-prey interactions the prey can escape in space by colonizing new environments, whereas the predator usually follows after some period of time. In this way, time delays in the response of the predator to prey density, which may cause instability at a single location, act to stabilize the dynamics in space. The same rules apply equally to unstable competitive interactions where the weaker competitors often have superior dispersal powers and are thus able to keep ahead of their more competitive rivals.
6. Space may act as a sieve to weed out the physiologically or genetically weaker migrants so that new environments are colonized by only the hardiest individuals. With time, however, these new colonies may deteriorate as the stronger individuals leave the overcrowded conditions and as predators, diseases, and pollutants accumulate. Thus, the quality of populations may change dramatically in space and time.
7. Climatic and topographic variations determine the spatial favorability of an organism's environment and, therefore, the distribution and levels of abundance of the

species. Plant and animal communities evolve within the physical framework set by these parameters. Thus, geographic regions can be stratified according to their favorability for different organisms, and this provides the manager with a framework for predicting the dynamics of particular species and communities.

Notes

5.1. For those interested in further readings on the subject of migration and dispersal in relation to environmental or habitat conditions, the work of T. R. E. Southwood is suggested as a starting place. In particular, his articles in *Biological Review* (vol. 37, p. 191, 1962) and in the *Journal of Animal Ecology* (vol. 46, p. 337, 1977) are recommended. In the latter paper, Southwood developed the idea of the habitat acting as a template for the evolution of life-cycle strategies, with migration and dormancy as the main tactics for dealing with environmental variations in time and space.

5.2. When a species is introduced into a favorable environment where it was not previously present, we are able to observe its spread through the new environment from the point of introduction. One can develop an analytical model of this expansion through space by assuming that dispersal is effectively random within a large geographic region. A thorough mathematical analysis of random dispersal was published in 1951 by J. G. Skellam in *Biometrika* (vol. 38, p. 196), in which he developed a theory of dispersal analogous to Brownian motion and the diffusion of gases. With the additional assumption that an organism invading a new environment has unrestricted population growth, Skellam showed that the rate of spread is approximately constant, being proportional to the maximum individual rate of increase and the dispersal powers of the organism, and that the area occupied by the population increases linearly with the square of the time from introduction. This result is analogous to the well-known Inverse Square law, and can be written

$$A - r_m a^2 t^2 \ or \ \sqrt{A} = at \sqrt{r},$$

where A is the area occupied, r_m is the maximum instantaneous rate of per capita increase, a is the coefficient of dispersal, and t is the time from introduction. Skellam found good correspondence between his theory and data for the spread of the muskrat in central Europe after its introduction in 1905. The theory has also been used to describe the spread of the larch casebearer in western North America by G. E. Long (*Environmental Entomology*, vol. 6, p. 843, 1977). In this case the relationship between $\sqrt{A}$ and t had three linear segments (see Figure 5.13) which Long interpreted as a period of adaptation to the new environment where the rate of spread was fairly low, a maximum rate of spread of the adapted insect, and a period of stabilization when the insect was approaching saturation of the area. Long also modeled the spread of a parasitic wasp, which was introduced to control the foliage-eating casebearer,

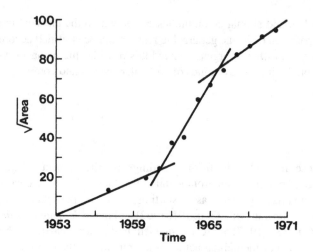

Fig. 5.13 The spread of the larch casebearer in northern Idaho (redrawn from Long; referenced in this note)

and came to some interesting conclusions concerning the biological control of the casebearer.

Skellam's theory provides a convenient analytical platform for evaluating the spread of a population from a point source. However, when movements are occurring from many different points in space the analysis becomes less tractable and we usually have to resort to simulation. On the other hand, systems such as the spruce budworm-forest interaction discussed in this chapter can be evaluated analytically by introducing Skellam's diffusion theory into the single compartment population model. This has been done by D. Ludwig, H. F. Weinberger, and D. Aronson, as reported by W. C. Clark, D. D. Jones, and C. S. Holling in the book *Spatial Pattern in Plankton Communities* (edited by J. S. Steele, Plenum Press, 1979). One of the most interesting results of this analysis was that a critical patch size was necessary for the spruce budworm population to persist; that is, the patches of favorable environment, mature balsam fir, must be greater than some critical size before budworm outbreaks can occur.

5.3. A description of the spruce budworm model can be found in the *Proceedings of a Conference on Pest Management* (edited by G. A. Norton and C. S. Holling, published by the International Institute of Applied Systems Analysis, Laxenburg, Austria, 1977). The population dynamics of the budworm, on which the model was based, is reported in detail in the *Memoirs of the Entomological Society of Canada*, no. 31, 1963 (R. F. Morris, editor).

5.4. For those interested in pursuing the theory of insect epidemiology, later developments are summarized and extended in the paper by the senior author of this book in *Researches in Population Ecology* (vol. 19, p. 181, 1978). In this paper he also discusses the roles of space and dispersal in the spread of epidemics and the concept of epidemic thresholds.

5.5. Kenneth Watt provides a succinct summary of the dynamics of epidemics in Chapter 6 of his book *Ecology and Resource Management* (McGraw-Hill Book Company, New York, 1968), which – together with the references cited therein – should form a good starting point for those interested in delving into the voluminous literature on the subject of disease epidemiology.

5.6. Because thresholds describe the boundary between two distinct system behaviors, they represent transient system states and as such cannot be directly observed. Thus, the direct empirical determination of an epidemic threshold is usually impossible. However, thresholds can sometimes be approximately located, as was done in Figure 5.6, by plotting data from a number of endemic and epidemic states in the phase space of the critical variables. The critical variables can often be separated into two sets, those that determine the potential epidemic behavior (e.g., stand conditions in the case of the spruce budworm), and those that enforce the endemic equilibrium by preventing this potential from being attained (e.g., the density of insectivorous birds). In the case of the bark beetle illustrated in Figure 5.6, the potential beetle population is determined by the number of thick-phloemed trees in the stand that are favorable for reproduction and survival. However, the beetle may be prevented from utilizing most of these trees if they are vigorous and can resist the beetle's attack. If these variables can be measured in a number of stands that contain endemic and epidemic beetle populations, then the position of the epidemic threshold can be approximated by drawing a line separating the two population behaviors.

For those who wish to pursue the subject of threshold theory, they are referred to the paper by the senior author of this book in *Researches in Population Ecology* (Note 5.4). Details of the ecology and epidemiology of the mountain pine beetle, and the derivation and application of threshold functions, can be found in the proceedings of a symposium on the *Theory and Practice of Mountain Pine Beetle Management in Lodgepole Pine Forests*, edited by A. A. Berryman, G. D. Amman, R. W. Stark, and D. L. Kibbee and published in 1978 by the Forest, Wildlife and Range Experiment Station of the University of Idaho, Moscow. A more general view of the theory of thresholds by this author was published in a book edited by G. R. Conway entitled *Pest and Pathogen Control: Strategy, Tactics, and Policy Models* (Wiley Interscience, New York, 1981).

5.7. The effect of predator movements on the stability of predator-prey interactions has also been evaluated by considering the tendency of predators to aggregate in areas where their prey are most abundant. This aggregation effect can be thought of as a response by the predator to the favorability of its environment: When food is scarce in a particular area the predator will move greater distances searching for prey and this will cause it to move away. When food is abundant, it will encounter prey more frequently after moving only short distances and, therefore, it will tend to remain in the same area. For obvious reasons, animals with learning abilities will have much stronger tendencies to remain in areas where their food is abundant.

The net effect of predator aggregations is to keep the predators in regions of dense prey populations and to accentuate the time delay in exploiting newly established prey populations. As we know, this will tend to stabilize the predator-prey system in space.

A number of theoretical models have been built to examine the effect of predator aggregations on the stability of predator-prey systems and they all show that stability is increased by such behaviors. For those interested in pursuing this subject, Michael P. Hassell presents a detailed review in his book *The Dynamics of Arthropod Predator-Prey Systems*, published by Princeton University Press, New Jersey, 1978.

5.8. One of the interesting results of Skellam's analysis of the dispersal process (see Note 5.2) was that a weak competitor can persist in space, even when outcompeted in a particular area, provided it has much stronger dispersal powers than its rivals. We reached this same conclusion in Chapter 4 when we discussed the strategy of the opportunist.

5.9. The theory of qualitative population changes in space was pioneered by W. G. Wellington in his studies of a forest defoliator, the western tent caterpillar. Wellington showed that the more vigorous individuals (phenotypes, or perhaps genotypes) were more prone to move out of overpopulated areas and that the remnants were more susceptible to predation and disease. These qualitative changes, operating within a highly variable environment, enabled the insect population to persist in space and time. Wellington and his colleagues also developed a spatially defined computer simulation model, much like the spruce budworm model, which produces scenarios of population buildup, expansion, deterioration, and collapse. For those interested in pursuing this fascinating study the following papers are recommended: W. G. Wellington, *Canadian Entomologist*, vol. 96, p. 436 (1964) and vol. 97, p. 1 (1965); W. G. Wellington, *Canadian Journal of Zoology*, vol. 38, p. 289 (1960); W. G. Wellington, P. J. Cameron, W. A. Thompson, I. B. Vertinsky, and A. S. Landsberg, *Researches on Population Ecology*, vol. 17, p. 1 (1975); and W. A. Thompson, P. J. Cameron, W. G. Wellington, and I. B. Vertinsky, *Researches on Population Ecology*, vol. 18, p. 1 (1976).

Data from other animal populations, such as lemmings and small rodents, seem to support Wellington's theory. In particular, vole populations exhibit dramatic genetic changes in response to population density as demonstrated by C. J. Krebs and J. H. Myers (*Advances in Ecological Research*, vol. 8, p. 267, 1974).

5.10. The concept of "habitat type" for classifying land according to its potential for supporting a particular vegetational climax association was first introduced by Rexford and Jean Daubenmire in the *Technical Bulletin of the Washington Agricultural Experiment Station* (no. 60, Washington State University, Pullman, 1968). Habitat-type classifications are now used extensively throughout the Rocky Mountain states as a basis for stratifying and managing public lands administered by the United States Forest Service.

The concept of habitat type is essentially the same as our ideas of environmental favorability. After all, a habitat is nothing more than a particular environment where a particular organism lives. The habitat type is, therefore, a measure of the favorability of that environment. Because we have developed our population theory around the more general concept of environment, we will attempt to use the term "habitat" rather sparingly (see also Note 3.7 for the synonymous term "habitat suitability" introduced by Fretwell).

5.11. One of the first, and perhaps most successful, risk classifications systems was designed by Paul Keen (Journal of Forestry, vol. 34, p. 919, 1936). Keen classified ponderosa pines according to their risk of attack by the western pine bark beetle. Although the original classification has been modified considerably with time, it is still used as the basis for ponderosa pine management in certain areas of California. High-risk trees, which are favorable environments for reproduction of the destructive bark beetle, are selectively harvested so that the beetle population rarely reaches destructive levels.

5.12. The risk classification map shown in Figure 5.11 was based on a system designed by G. D. Amman, M. D. McGregor, D. B. Cahill, and W. H. Klein (U.S. Department of Agriculture Forest Service Technical Report INT-36, 1977). In this system risk categories are assigned according to the age, mean diameter, and elevation of the stand of lodgepole pines; older, large diameter stands growing at low elevations being at greatest risk. More recent studies reported by R. L. Mahoney and A.A. Berryman in the book *Theory and Practice of Mountain Pine Beetle Management in Lodgepole Pine Forests* (see Note 5.6 for complete reference) suggest that this classification system can be improved by including information on the vigor of the stand; that is, its density and periodic growth rate. In this way the position of the stand relative to the outbreak threshold can be estimated (see Note 5.6 and Figure 5.6).

5.13. Within the aphids (Hemiptera: Homoptera), polyphenism, i.e., environmentally determined phenotypic differences, is one of the key factors determining their importance as pest species. Wing (alary) polyphenism and polyphenism associated with the mode of reproduction are the most widely known. Wing polyphenism is essential for the aphid life cycle and, by allowing migration to fresh resources, it may contribute to determining the overall fitness of an aphid clone. Aphids have complex life cycles but during the summer most species reproduce asexually and live as clonal, fast-growing colonies. Depending on environmental cues such as day length and temperature, these colonies produce sexual morphs in the autumn when mating occurs. Ten per cent of aphid species are host alternating (heteroecious), moving between woody and herbaceous host plants according to season. Migration between the two different species of host plant clearly requires winged morph production by the colony. Even for non host-alternating (autoecious) species, however, wing induction is important, as winged morphs allow a clone to take advantage of several individual host plants during a season when the quantity and quality of the host plants change. Thus winged morph production in aphids is a phenotypic trait that has traditionally been seen as a response to unfavorable environmental conditions: when aphids destroy the host plant, or when the host plant quality deteriorates for other reason (e.g., the amount of nutrients declines in trees during the summer) they produce winged morphs and fly away to find another, more suitable host. For those interested in details of this phenomenon, we recommend the paper by C. B. Muller et al.: The role of nutrition, crowding and interspecific interactions in the development of winged aphids, published in Ecological Entomology, Vol. 26, pp.330–340, in 201.

Chapter 6
Interactions Between Many Species (Ecological Communities)

Community ecology is a discipline that deals with the interrelationships between assemblages of plant and animal populations that live together at a particular time and in a particular place. A large body of information and theory has developed around this branch of ecology, which would be sufficient for a book in its own right. Therefore, it is not the purpose of this chapter to present an exhaustive treatment of community ecology, but rather to form a link between our concepts of population dynamics and those of community ecology.

6.1. Community Structure

The assemblages of plant and animal populations that make up ecological communities often possess well-defined spatial boundaries, which separate them from other communities. These boundaries can usually be recognized by rather abrupt changes in the dominant species in the community (usually plants but sometimes animals, as in coral reefs), or in the physiographic structure of the landscape. Thus, we recognize a grassland community from an oak-hickory community, a coral reef from a sandy bottom community, and so on.

The characteristic fauna and flora, which make up a typical community, interact with each other as cooperators (mutualism or symbiosis), as competitors for common resources, and as predator and prey. The interaction network linking the different species is usually called a *food web* or *food chain*, indicating that the primary interactions are over food, either through eating one another or competing or cooperating with one another for food resources. At the base of all food chains are the producer organisms, which synthesize carbohydrate and protein from raw materials. Plants, of course, perform this function through the basic process of photosynthesis, where the raw materials – water, carbon dioxide, and energy in the form of sunlight – are combined to produce carbohydrates, the basic building blocks of all biological organisms. Other raw materials, in the form of mineral nutrients and salts (nitrates and phosphates in particular), are used to create amino acids and their protein derivatives.

Herbivores, which feed on the producer organisms, have the same problems as all predators in maximizing their own numbers while practicing conservation. Because

A.A. Berryman, P. Kindlmann, *Population Systems: A General Introduction*
© Springer Science+Business Media B.V. 2008

of the requirements of conservation, and because conversion of plant biomass into herbivore biomass is rather inefficient, the standing crop of herbivores must be considerably less than that of the food plants. For the same reason, carnivore biomass will be much less than that of its herbivorous food. Thus, we find a pyramid of numbers, or more correctly biomass or standing crop, such as that in Figure 6.1, in most ecological communities. It is evident from this diagram that food chains have finite length, which is dependent on the efficiency of energy conversion between trophic levels. In fact, we will rarely find food chains with more than three or four trophic levels; for example, plants, herbivorous insects, parasitoids, and hyperparasitoids. Occasionally food chains with five trophic levels can be found, particularly in the ocean where the standing crop of producers is extremely large; for example, phytoplankton, zooplankton, fish, seals, killer whale.

Each trophic level in an ecological community will be composed of one or more species, which may compete, or sometimes cooperate, with each other for resources. The overall interaction network can be represented by a community web, or interaction matrix, such as that illustrated in Figure 6.2. The community structure is defined by the signs of the interaction effects and the feedback loops that they create. We can see that, in addition to the competitive (− −), predatory (+ −), and other feedback

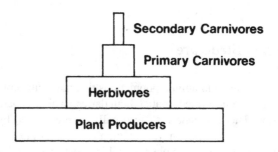

Fig. 6.1 A trophic pyramid of biomass or standing crop

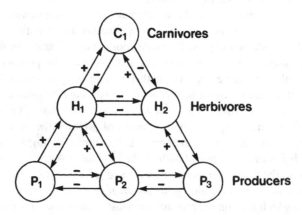

Fig. 6.2 The structure of a simple community composed of three producers, two herbivores, and a carnivore

loops that link two species together, we also have feedback loops that involve three or more species. For example, in Figure 6.2 we have the feedback loop, $C_1 \xrightarrow{-} H_1 \xrightarrow{-} H_2 \xrightarrow{+} C_1$, which involves a carnivore and two herbivores. This feedback loop, which has two negative and one positive link, has an overall positive feedback effect, which we would expect to cause instability in the system. On the other hand, the four species loop, $E_a \xrightarrow{+} N_b \xrightarrow{+} E_b \xrightarrow{+} N_a \xrightarrow{+} E_a$. has an overall negative feedback effect, which should help to stabilize the system. An important question, therefore, is whether the combination of all these loops produces a system that is stable or unstable and, if it is unstable, what adjustments in the structure are necessary to create a stable community. We will address these questions in the next section of this chapter.

6.2. Community Stability

The stability of an ecological community is determined by the properties of the individual populations and the network of interactions linking these populations together. When the number of species present in the community, and their relative abundance, remains fairly constant in time, then the community is considered to be stable. Because communities are subject to unpredictable variations in their physical environments, a stable community is also one that can recover its characteristic composition and relative abundances following an environmental disturbance. That is, it is *resilient* to disruptive influences, whether they are natural or man-made (see Chapter 3 and Note 3.9 for a discussion of resilience).

One of the central tenets of classical ecology is that complex communities tend to be more stable than simple ones. This doctrine was based largely on the observation that the complex communities of tropical regions are, on the whole, more stable in time than the simple communities of temperate regions, which are often characterized by large-scale population fluctuations, pest outbreaks, and the like. In the seventies, however, this belief has been challenged by mathematical arguments that suggest that systems made up of complex interaction networks are less stable than simple ones (see Note 6.1). In fact, Robert May presented the contrary opinion that stable communities may become more complex because they are less subject to external disturbances; that is, stability permits complexity rather than the other way around as proposed by classical theory. In this chapter we will concern ourselves less with the problem of complexity versus stability and more with the conditions that are required for organisms to coexist in ecological communities.

Because communities may be composed of a large number of species and have an extensive interaction network, we need to develop a less complicated model for the individual population system. We can do this by reducing the details of the population system into a single state description and relegate the feedback structure to a single loop. When species are combined to form communities we will obtain a feedback diagram with self-loops, S's, feeding back to the individual populations and interaction loops, C's, linking them together (Figure 6.3). Each circle, or *node*, in this diagram represents the state of a particular population, usually its density or biomass, and the arrows represent the direction of the interactions (note that a different convention was used in our block diagrams where boxes represented interaction processes, and arrows, population state variables). The impact of each interaction is specified by a parameter, whose sign determines whether the effect is positive or negative. A feedback loop is defined as an interaction or a series of interactions that eventually return to the starting node. As we know, the overall effect of a feedback loop is the product of all its interaction effects. Thus, the feedback between A and B in Figure 6.3 is $(- C_{ab})(- C_{ba})$, which we recognize as the positive feedback loop $C_{ab} C_{ba}$ caused by competition between the two species for common resources. Therefore, the system defined in Figure 6.3 is analogous to the more complicated two-species competition model of Chapter 4.

The general qualitative stability of feedback systems can be evaluated using the techniques of *loop analysis* (see Note 6.2). Loop analysis involves some rather formidable mathematics, so we have attempted to abstract a less formal, and hopefully more intuitive, version for the purposes of this book. To perform a loop analysis we first have to assume that the system is at or near equilibrium. In other words, we will perform a steady-state analysis on the system in the neighborhood of its equilibrium point, and ask the question "Will the system return to this equilibrium following a small displacement from it?" From our knowledge of general systems theory we know that equilibrium can only be stable if it is dominated by negative feedback. The same principle holds in loop analysis; that is, the total community is stable if, and only if, negative feedback dominates at all levels of organization. The levels of organization identify those feedback loops that involve one, two, three, etc., species. Hence, in Figure 6.3 two feedback loops involve only one species and these are S_{aa} and S_{bb}. At the second level we have the loop between both species C_{ab} C_{ba} and, in addition, the two separate loops S_{aa} and S_{bb} because, once again, two species are involved. Loops that are present at a level of organization but do not share common nodes are called separate or *disjunct* loops, while those that share nodes are joined or *conjunct* loops. Thus, Figure 6.3 contains one conjunct and one disjunct loop configuration at the second level of organization.

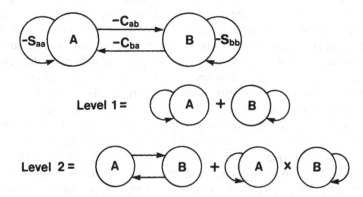

Fig. 6.3 A simple community composed of two species that are self-regulating and compete with each other for common resources, showing loop analysis at level 1 and 2

When the sum of the feedback loops at a particular level is negative, then the system is said to be qualitatively stable at that level of organization. For the total system to be stable, the sums of their feedback loops at all levels of organization must be negative.

Let us now evaluate the qualitative stability of the system depicted in Figure 6.3. The total feedback at level 1 is

$$F_1 = \Sigma S_{ij}$$
$$= (-S_{aa}) + (-S_{bb}) = -(S_{aa} + S_{bb}) \tag{6.1}$$

and the system is stable at this level of organization. At level 2 we have a conjunct loop plus the product of the disjunct loops, so that

$$F_2 = (C_{ab}C_{ba}) + (-S_{aa})(-S_{bb}).$$

However, the rules of loop analysis insist that, when all disjunct loops are negative, then their effect on the feedback at that level must also be negative. Therefore we have to change the sign (see Note 6.3) so that

$$F_2 = \Sigma C_{ij}C_{ji} - \Sigma S_{ii}S_{jj}$$
$$= C_{ab}C_{ba} - S_{aa}S_{bb} \tag{6.2}$$

The total feedback at level 2 is ambiguous because we have a negative and a positive term. This means that, in order to answer questions concerning the overall stability of the system, we need to know the magnitudes of the parameters. For example, it is easy to see that the community is stable if

$$S_{aa}S_{bb} > C_{ab}C_{ba},$$

because then F_2 will be negative.

This example illustrates the power of qualitative loop analysis. First, it is possible to determine the stability properties of a complex community knowing nothing about the magnitude of the many parameters. All we have to know is their signs, a property which is often obvious from the type of interaction; e.g., predation $(+ -)$, competition $(- -)$, commensalism $(+ 0)$, etc. Secondly, the analysis may identify which of the parameters needs to be measured in order to resolve ambiguous results at any particular level of organization.

Let us now look at a slightly more complicated community composed of three competing species (Figure 6.4). Feedback at level 1 will equal the sum of the three self-loops [equation (6.1)]

$$F_1 = -(S_{aa} + S_{bb} + S_{cc})$$

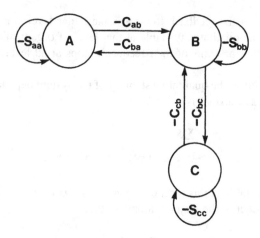

Fig. 6.4 A simple community of three self-regulated species competing for resources, where B competes with both species but A and C do not compete

and the system is stable at this level of organization. At level 2 we now have two conjunct loops linking A to B and B to C, and three disjunct loops made up of the self-loops A and B, B and C, and C and A [equation (6.2)]. Hence,

$$F_2 = (-C_{ab})(-C_{ba}) + (-C_{bc})(-C_{cb}) - (-S_{aa})(-S_{bb}) - (-S_{bb})(-S_{cc}) - (-S_{cc})(-S_{aa})$$
$$= C_{ab}C_{ba} + C_{bc}C_{cb} - S_{aa}S_{bb} - S_{bb}S_{cc} - S_{cc}S_{aa}$$

Once again feedback at level 2 is ambiguous and the system can only be stable if the sum of the self-loop products is larger than the sum of the interaction loops; that is

$$S_{aa}S_{bb} + S_{bb}S_{cc} + S_{cc}S_{aa} > C_{ab}C_{ba} + C_{cb}C_{bc}.$$

Because stability cannot be determined at level 2 without information on the magnitudes of the parameters, there is really no need to proceed to an analysis of feedback at level 3. One of the basic rules of loop analysis is that instability due to positive feedback at one level cannot be corrected by negative feedback at a higher level of organization. However, in order to illustrate the procedure for level-3 analysis, we will continue. The general equation for feedback at this level is

$$F_3 = \Sigma C_{ij}C_{jk}C_{ki} - \Sigma S_{ij}C_{jk}C_{kj} + \Sigma S_{ii}S_{jj}S_{kk}$$

(6.3)

(see Note 6.3 for the derivation of the equation). The first expression of this equation specifies the sum (Σ) of all the single loops that pass through all three nodes; that is single loops involving A, B, and C. As A and C do not interact in this example, there are no single loops of length three. The second term is the sum of the products of disjunct loops involving a self-loop and a two-species interaction loop. In our example we have two of these, $(-S_{aa})(C_{bc}C_{cb})$ and $(-S_{cc})(C_{ab}C_{ba})$. There is no disjunct loop involving S_{bb} because species B is a component of both two-species interactions. The third term of the equation is the sum of the products of all combinations of three self-loops. In the example there is only one of these, $(-S_{aa})(-S_{bb})(-S_{cc})$. Thus, total feedback at level 3 is

$$F_3 = 0 - (-S_{aa}C_{bc}C_{cb} - S_{cc}C_{ab}C_{ba}) + (-S_{aa}S_{bb}S_{cc})$$
$$= S_{aa}C_{bc}C_{cb} + S_{cc}C_{ab}C_{ba} - S_{aa}S_{bb}S_{cc}$$

and, once again we see that the result is ambiguous. However, S_{aa} and S_{cc} now contribute to the positive feedback components, while at level 2 they only contributed to negative feedback. Hence, the constraints on stability are even more restrictive at level 3 because now S_{aa} and S_{cc} must not be too large. This result emphasizes the rule that stability will never be increased, and will usually be decreased, by feedback at higher levels of organization.

It is interesting to note that the system illustrated in Figure 6.4 can be stable if the self-limiting feedback acting on species B is very strong relative to the other interactions. This is because the parameter S_{bb} is the only one that contributes solely to negative feedback at all three levels of organization. In other words species B must be close to its carrying capacity, where intraspecific competition is most intense. For this to be possible the effects of the two competitors on this species must be correspondingly weak; that is, the parameters C_{ab} and C_{cb} must be small. We would probably arrive at the same conclusion intuitively because species B, having to deal with two competitors, is under much more pressure to evolve ways in which to avoid competition. It could do this by becoming a strong competitor or by adopting an opportunistic life style.

We can continue to make the community more complicated by adding more competing species, or more interactions between them, and we would find that the requirements for stability become more and more restrictive as we make the community more complex (the student is encouraged to try this). The main result that emerges from such exercises is that the interactions between competing species must be much weaker than the self-regulating mechanisms if the populations are to persist in the community. In other words, as more species enter an ecological community they will have to evolve ways in which to reduce or eliminate competitive interactions if the community as a whole is to remain in equilibrium. From the other side, we see that competitive interactions are often the driving forces in the dynamics of community succession and that a state of relative stability is only attained by the climax community. Thus, it is the climax species that have evolved ways in which to live with their competitors by reducing their competitive interactions.

Let us now turn our attention to communities involving higher trophic levels; that is, plant-herbivore and herbivore-carnivore interactions. First consider a simple community composed of a self-regulated prey and a predator that is only limited by the abundance of its prey (Figure 6.5).

As there is only one self-loop in this system, the feedback at the first level of organization is defined by

$$F_1 = -S_{pp}.$$

At the second level we have a single conjunct loop, the predator-prey interaction, and no disjunct loops, and so the total feedback is

$$F_2 = (C_{pa})(-C_{ap}) = -C_{pa}C_{ap}.$$

Thus, the community is stable provided that there are no long time delays in the numerical responses of predator or prey. Now let us complicate this system by adding another predator, which competes for the same prey (Figure 6.6). Feedbacks at the first and second levels are

$$F_1 = -S_{pp}$$
$$F_2 = (C_{pa})(-C_{ap}) + (C_{pb})(-C_{bp}) + (-C_{ab})(-C_{ba})$$
$$= -C_{pa}C_{ap} - C_{pb}C_{bp} + C_{ab}C_{ba}$$

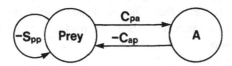

Fig. 6.5 A simple community consisting of a predator A feeding on a self-regulated prey

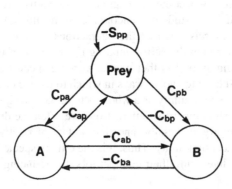

Fig. 6.6 A simple community composed of two predators, A and B, feeding on a common self-regulated prey

which means that the community will be stable provided the positive feedback between competing predators is weaker than the sum of the interactions with the prey. However, even if these conditions are met, we find that the system is completely unstable at level 3. Feedback at this level is made up of two loops which involve all three species, namely the loop $(-C_{ab})(-C_{bp})(C_{pa})$ and its reverse $(-C_{ap})(C_{pb})(-C_{ba})$, both of which are positive, plus a single disjunct loop composed of the prey's self-loop and the competitive interaction between the predators, which also turns out to be positive. Hence,

$$F_3 = C_{ab}C_{bp}C_{pa} + C_{ap}C_{pb}C_{ba} + S_{pp}C_{ab}C_{ba}$$

and the community is unstable whether the conditions for stability at level 2 are met or not.

This result is not really surprising because we have a case of two species competing for exactly the same resource, their common prey. In such cases the competitive interactions must be extremely strong and competitive exclusion is the most likely result – this is aptly demonstrated by the unstable community of parasitoids feeding on a common prey shown in Figure 4.7A. This system becomes much more stable, if the two predators limit their own numbers, say by territorial behavior. The student is invited to prove this.

There are many examples, however, of prey populations that are fed upon by a complex of predators and that seem to persist in a stable natural community. How then can we justify such observations with the results of our loop analysis? The first, and perhaps most obvious observation, is that different predators often stratify their attacks on a common prey in both space and time; that is, they tend to attack their prey at different times and in different places so that they do not compete directly with each other. When this occurs the competitive interaction is broken and it is easy to show that the system is now stable ($F_1 = -S_{pp}$; $F_2 = -C_{pa}C_{ap} - C_{pb}C_{bp}$). Another feature of some predators, particularly invertebrates, is for them to feed on each other as well as on their common prey. If one of the predators is more successful at attacking its rival, then it can be considered a predator of both other species and the system will become much more stable. For instance, if species A (Figure 6.6) is the superpredator, then the interaction C_{ba} will be positive and the feedback structure becomes

$$F_1 = -S_{pp}$$
$$F_2 = -C_{pa}C_{ap} - C_{pb}C_{bp} - C_{ab}C_{ba}$$
$$F_3 = C_{pa}C_{ab}C_{bp} - C_{pb}C_{ba}C_{ap} - S_{pp}C_{ab}C_{ba}$$

This system will be stable as long as the loop $C_{pa}\,C_{ab}\,C_{bp}$ is not too strong. One of the interesting facets of this result is that stability can be improved if species B reduces its impact on the common prey; for instance by assuming a scavenging life style, in which case $C_{bp} = 0$ and the positive loop in the level 3 feedback is eliminated.

At this point it is necessary to pause for a moment and consider again the problem of time delays in our negative feedback structure. As negative feedback loops become longer, or involve more and more species, the possibility of delays creeping in becomes more likely. As we know, these delays can cause the system to oscillate in an unstable manner around the equilibrium position (see Chapter 2). In general, therefore, it is necessary for negative feedback at the higher levels to be weaker than that at the lower levels before the system is considered stable. The criterion is that

$$F_1 F_2 + F_3 > 0.$$

$$(6.4)$$

In the case in question, where species A is a predator on both common prey and its rival (see the equations above), then this criterion becomes

$$S_{pp} C_{pa} C_{ap} + S_{pp} C_{pb} C_{bp} + C_{pa} C_{ab} C_{pb} - C_{pb} C_{ba} C_{ap} > 0,$$

which should hold under almost all imaginable conditions. In the case where species B becomes a scavenger, the stability criterion can be shown to be (the student is encouraged to do this)

$$S_{pp} C_{pa} C_{ap} - C_{pb} C_{ba} C_{ap} > 0$$

in which case oscillatory instability is possible if the interactions in the long feedback loop are too strong. However, the system will be stable if the prey has strong self-regulatory effects.

The next plausible evolutionary step in this community is for the super-predator to disengage its interaction with the common prey and feed entirely on the other predator. This adds another trophic level to the community and produces a much more stable system (the student is encouraged to show that $F_1 = - S_{pp}$, $F_2 = - C_{bp}$ $C_{pb} - C_{ab} C_{ba}$, $F_3 = - S_{pp} C_{ab} C_{ba}$, and $F_1 F_2 + F_3 > 0$).

The above arguments lead to some interesting speculations concerning the evolution of ecological communities. For instance, if we have an unstable community consisting of two predators feeding on a common prey, then there are two major evolutionary pathways to a stable community: First, the predators can adapt to feed at different times or places, or to attack different sizes or species of prey, in which case direct competition will be minimized or eliminated. Second, one of the predators can adapt to feed on its rival as well as on the common prey. This community can become even more stable if the inferior predator also adapts to reduce its impact on the prey by assuming a scavenging or parasitic life style. The most stable three-species community would be created if the superior predator adapted to feed upon the other predator alone.

Because stability is a necessary criterion for the persistence of an ecological community, then the requirements for stability can be viewed as selective forces, which lead to the evolution of specialized behavior, which in turn modify the interspecies interactions to create a stable system. This process can be viewed as an evolutionary feedback loop, which operates in the following way: Given an unstable community, then one or more species will be driven toward extinction. This puts selective pressure on them to modify their interactions with the other members of the community, and these adaptations change the feedback structure and the stability properties of the system. Those species that are successful in adapting will, of course, retain their place in the community while the unsuccessful will disappear. The communities that we observe in nature are the outcome of this evolutionary game and, therefore, will usually be composed of a mixture of co-evolved species, which have "learned" to live with each other in relative harmony.

If we accept the proposition that stable communities arise as a result of selective forces molding the genetic properties of individual species to produce stable feedback loops (self-limiting and trophic interactions) and to weaken or break unstable ones (competitive interactions), then we can argue that complex communities can only be formed as a result of this co-evolutionary process. In other words, complex communities can only evolve if each component species has the *evolutionary time* to adapt to all the species with which it interacts. However, because the interspecific interactions are also affected by the physical environment (e.g., temperature, precipitation, soil structure, and nutrients, etc.), it is difficult to see how this co-evolutionary process can proceed in a changeable physical setting: How can a species adapt to interactions that are constantly changing? It seems, therefore, that complex communities will only be able to evolve in rather constant physical environments (see also Note 6.4). In very variable physical environments there will be little time available for interspecific adjustments to evolve, and we would expect to find much simpler ecological communities. This argument leads us to the general proposition that environmental stability permits the evolution of complex communities. There is an important lesson in this conclusion which man is slowly learning from experience. If we expose these complex communities to more variable conditions, such as human agricultural practices, then we can upset the delicately co-evolved structure and possibly create an unstable system. From this perspective, complex systems seem to be much more fragile than simple ones, or they are much less resilient to changes imposed from outside (see Note 3.9).

6.2.1. Predation as a Stabilizing Influence

The question of predators acting as a stabilizing influence on otherwise unstable competitive interactions has intrigued many community ecologists. The question arose originally from experimental studies in which predators were excluded from a community, and subsequent observations, which revealed that complexity decreased as competing species disappeared from the community (see Note 6.5). These experiments suggested that predators can indeed stabilize communities and, thereby, increase the richness and diversity of ecological systems. Let us examine this question using a simple one predator-two prey system (Figure 6.7). The feedback structure is defined by

$$F_1 = -S_{aa} - S_{bb}$$
$$F_2 = C_{ab}C_{ba} - C_{ap}C_{pa} - C_{bp}C_{pb} - S_{aa}S_{bb}$$
$$F_3 = C_{ab}C_{bp}C_{pa} + C_{ap}C_{pb}C_{ba} - S_{aa}C_{bp}C_{pb} - S_{bb}C_{ap}C_{pa}$$

The system is indeterminate at levels 2 and 3. However, the presence of the predator has added two extra negative terms to feedback at level 2 (compare this with Figure 6.3, which specifies feedback with two competing species). In addition, feedback at level 2 can be negative even when the competitive interactions are stronger than the competitors' self-limiting effects, provided that the predatory loops are strong; that is,

$$C_{ab}C_{ba} - S_{aa}S_{bb} < C_{ap}C_{pa} + C_{bp}C_{pb}.$$

We would expect this condition to hold in most cases where the predator is fairly efficient or the prey fairly vulnerable to attack. However, the constraints at level 3 are much more restrictive. Here the only interactions that do not appear in the negative terms are competitive, and the only ones that do not appear in the positive side are self-loops. Therefore, stability at level 3 seems unlikely when the competitive interactions are stronger than the self-regulatory effects. We can demonstrate this by assuming that the predator feeds equally on both prey species so that $C_{ap} = C_{bp} = C_{.p}$ and $C_{pa} = C_{pb} = C_{p.}$, and then factor out the predatory interactions to give

$$F_3 = C_{.p}C_{p.}(C_{ab} + C_{ba} - S_{aa} - S_{bb}).$$

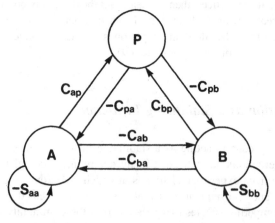

Fig. 6.7 A simple community consisting of a general predator, P, feeding on two competing prey species. A and B

We can see that negative feedback only dominates when the sum of the self-loops is greater than the sum of the competitive interactions (i.e., $S_{aa} + S_{bb} > C_{ab} + C_{ba}$).

So far our analysis suggests that general predators cannot stabilize competitive interactions that are themselves unstable. How then can we rationalize the experimental evidence, which support the opposite contention? One possible explanation is that the predator maintains the prey at such low densities that competition between them is for all intents and purposes eliminated. This may occur if the predator is extremely efficient, or the prey very vulnerable, and if the intensity of competition between the prey is directly related to their densities (i.e., the case illustrated by Figure 4.9B). In this case, competition will become weaker and weaker as the equilibrium prey densities decline under predation, and we may find a stable community with $C_{ab} + C_{ba} < S_{aa} + S_{bb}$. However, if the predator is removed, then the prey populations will grow and the increasing competition may create the unstable situation $C_{ab} + C_{ba} > S_{aa} + S_{bb}$.

Another plausible explanation involves the tendency for some predators to switch their feeding preferences to the more abundant prey species (see Chapter 4). When this happens their interaction with the scarce prey species weakens and the negative feedback at level 3 will increase, creating a more stable community. For example, suppose that prey B (Figure 6.7) is the most abundant species at a particular time and that the predator concentrates its feeding on this population. The interactions between the predator and the scarce species will become very weak and this will cause both positive terms of level-3 feedback to decrease while only one negative term decreases. In particular, we can see that when $C_{ap} = C_{pa} = 0$, then $F_3 = - S_{aa} C_{bp} C_{pb}$ and the community is stable. We must conclude, therefore, that predators can indeed stabilize competing communities, which would otherwise be unstable, if they hold the density of their prey populations to levels where interspecific competition is relatively weak, or if they preferentially crop the more abundant species. In this way, predation may increase the diversity of ecological communities and permit greater complexity.

We can, of course, continue to ask questions concerning the stability of real or imagined communities of varying degrees of complexity. However, knowing the rules of loop analysis, the student can pursue such interesting games by himself (some examples of specific communities are provided for the student in the exercise section of this chapter). In general, however, we will usually find that the rules of two-species interactions hold in more complex associations, and that competitive and cooperative interactions tend to destabilize communities while trophic interactions tend to stabilize them.

6.3. Community Dynamics

Although ecological communities may evolve toward stable structures, given enough time, this process can be disrupted by catastrophic environmental forces (fires, volcanic eruptions, tornados, etc.) as well as by the activities of man (timber and animal harvesting, clearing agricultural lands, pollution, etc.). Following such catastrophic events, the disturbed area is usually invaded by pioneer plant species, the opportunists with their strong dispersal abilities or other adaptations, which take advantage of changing environments. For example, a proportion of the cones of lodgepole pine only open to release their seeds when subjected to intense heat, giving this species an advantage in re-colonizing areas denuded of life by forest fires. Pioneer plants and their associated animal fauna are usually the first communities to appear on disturbed sites. However, as we noted in Chapter 4, these species often create conditions that are unfavorable to their own reproduction and survival; that is, they form dense stands under which their offspring cannot survive because they are adapted to growing on open sites and, thus, are intolerant of shaded conditions. In time other shade-tolerant plants become established in the understory and will eventually replace the pioneer species. This succession of changing plant life is accompanied by a succession of animal species adapted to feeding on the different flora, and by a changing complex of predaceous species. Hence, we will observe a continually evolving community which, given sufficient time in an undisturbed state, will eventually stabilize as a climax community. This evolutionary climax association will usually be composed of shade-tolerant plants because only they can reproduce beneath their own canopies, and so individual plants that die tend to be replaced by their own kind.

In any given area, the climax plant community will be composed of species that are well adapted to the physiography (soil and topography) and climate of that region. These species are able to outcompete their less well-adapted rivals and so they and their animal associates eventually dominate the area. For this reason, climax communities will usually be separated from different neighboring communities by rather distinct boundaries, which will usually fall along topographic, edaphic (soil), and/or climatic discontinuities. However, there will usually be some overlap at the boundaries because environmental conditions are not too well defined and may become favorable for one group of species at one point in time or space and unfavorable at others. At the boundary, therefore, we often find transitional communities composed of species from both neighboring communities as well as others peculiar to the transition zone. For example, ponderosa pine outcompetes Jeffrey pine on basaltic soils while the reverse is true on serpentine soils, but at the boundary of these two soil types both species coexist (Note 6.6). Climax communities, therefore, form distinctive patterns in space and, superimposed on this, are the successional communities, which create a pattern that changes in time.

In regions where environmental disturbances occur at relatively frequent intervals, succession is continually disrupted and we may find pioneer communities succeeding each other. Thus, repeated fires in the Rocky Mountains encouraged the regeneration of lodgepole pine forests over vast areas of land. Nowadays, with modern fire control technology, many of these stands are being slowly converted to more shade-tolerant communities composed of Douglas-fir, true firs, and spruce. This succession is being hastened by the mountain pine bark beetle, which thrives in the older lodgepole pine stands and has devastated extensive areas of lodgepole pine forest in recent years (see Note 6.7).

Obviously, herbivores play a role in plant succession and the dynamics of the communities in which they live. We have seen that bark beetles remove mature and overmature pines a long time before they would normally die of old age or disease. Thus, the beetle, acting as nature's harvester, hastens the succession towards the climax association. This interaction becomes even more intriguing when we observe that the beetle usually removes the pines close to the time when their growth rates begin to decline (Note 6.7), or shortly after the stand has reached maximum productivity. Similar observations with other forest insects, notably the spruce budworm (see Chapter 5), have led to the proposition that herbivores actually regulate plant productivity close to its maximum for a particular site (Note 6.8). In this way, of course, they also tend to maximize their own productivity. Herbivores achieve this by feeding upon old, decadent, and sick plants, which creates more space for healthy, young competitors, which in turn increases plant growth rates as well as the productivity of the entire community. Acceptance of this hypothesis leads to the further proposition that community interaction networks have evolved so as to maximize productivity, or the rate of accumulation of biomass, rather than to maximize the standing crop (i.e., the total biomass). Let us examine this proposition by constructing the feedback system shown in Figure 6.8. Here we have divided the plant subsystem into two components, total biomass and productivity. As productivity must have a maximum we allow it to be self-limited. We have further assumed, in accordance with our hypothesis, that herbivore biomass is inversely affected by plant productivity; that is, herbivore populations increase when plant productivity declines. Performing a loop analysis we obtain

$$F_1 = -S_{pp}$$
$$F_2 = -C_{pc}C_{cp} - C_{hc}C_{ch}$$
$$F_3 = -C_{ph}C_{hc}C_{cp} - S_{pp}C_{hc}C_{ch}$$
$$F_1F_2 + F_3 = S_{pp}C_{pc}C_{cp} - C_{ph}C_{hc}C_{cp}$$

Hence the system is stable, although oscillatory instability is possible if the loop $C_{ph} C_{bc} C_{cp} > S_{pp} C_{pc} C_{cp}$; this may happen if plant productivity strongly suppresses herbivore populations (strong C_{ph} effect), as seems to be the case with

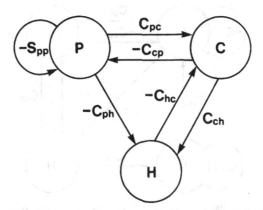

Fig. 6.8 A system composed of an herbivore, H, feeding on plants that have been split into the standing crop, C, and productivity, P, where productivity has a negative effect on the herbivore

the bark beetles and the spruce budworm. Thus, we might expect these species to exhibit cyclic growth and collapse patterns. Even though outbreaks of these species do occur, resulting in severe short-term losses to the standing crop, we see, however, that the *productivity* of the community is maximized *over the long run*.

We can also argue that the herbivores recycle nutrients from the sick and unproductive components of the community to the more productive, healthy individuals. For instance, the defoliation of forest trees causes increased litter accumulation, and the feces and decaying bodies of the herbivores release nutrients into the soil. Many herbivorous species, particularly insects, have much shorter generation spans than their plant hosts and, therefore, they increase the rate of nutrient recycling or, if you prefer, the turnover rate. If we incorporate the nutrient component in our feedback diagram we obtain an even more intriguing ecosystem (Figure 6.9). Loop analysis of this system gives us the feedback structure

$$F_1 = -S_{pp} - S_{nn}$$
$$F_2 = -C_{pc}C_{cp} - C_{hc}C_{ch} - S_{pp}S_{nn}$$
$$F_3 = -C_{ph}C_{hc}C_{cp} - C_{hn}C_{np}C_{ph} - S_{pp}C_{hc}C_{ch} - S_{nn}C_{pc}C_{cp} - S_{nn}C_{hc}C_{ch}$$
$$F_4 = C_{hn}C_{np}C_{pc}C_{ch} - S_{nn}C_{ph}C_{hc}C_{cp} - S_{pp}S_{nn}C_{hc}C_{ch}$$

which has negative loops at all levels of organization, *except* the loop $C_{hn} C_{np} C_{pc} C_{ch}$ at level 4. It is this loop that is so intriguing because it is composed of four positive interactions, $H \xrightarrow{+} N \xrightarrow{+} P \xrightarrow{+} C \xrightarrow{+} H$, which operate as a mutually beneficial positive feedback loop. In other words, the herbivore and plant act together as a *mutualistic* or *symbiotic* system at the fourth level of organization. Should this loop dominate the feedback at level 4, then the system will exhibit

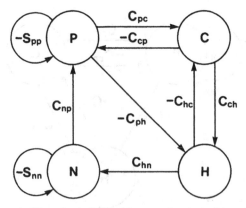

Fig. 6.9 An ecosystem similar to that in Figure 6.8 except that the herbivore increases the nutrient pool, N, which in turn acts to increase plant productivity

unstable self-enhancing growth; that is, plant productivity and biomass, as well as the herbivore population, will continually increase until negative feedback re-exerts dominance, as it must do eventually because there are limits to the nutrient pool and to plant productivity. This result is very important because it implies that predator–prey interactions may have mutualistic effects at the community level, actually improving the conditions for its prey rather than making them less favorable as we supposed until now. In addition, this result forces us to re-examine our thinking about pest species which may, in fact, be helping rather than hindering our efforts to maximize plant productivity, at least on a long-term basis (see also Note 6.8).

Of course, we can extend the above arguments to carnivorous species preying on the herbivores with similar results. However, if we add a carnivore to the system depicted in Figure 6.9 we will observe an even stronger mutually enhancing feedback loop at level 5 (the student is encouraged to demonstrate this by allowing the carnivore to contribute to the nutrient pool).

We have made but a brief excursion into the fascinating subject of community ecology and have skirted many interesting topics: for instance, the evaluation of community diversity, island biogeography and the problem of invasion and extinction of species, and the application of loop analysis and complementary feedback to evaluating evolution within ecological communities (see Notes 6.1 and 6.2 for further readings in these areas). However, we have tried to explore those areas where the evaluation of single-species or two-species models may lead to dangerous conclusions when applied to ecological communities, we have seen that most of the results we obtained with one- and two-species populations hold at the community level, but that communities possess some additional properties of their own. These, and the possibility of other undiscovered properties, should be on our minds as we play the role of manager in ecological settings.

6.4. Chapter Summary

In this chapter we have taken a brief look at the structure, stability and dynamics of ecological communities. The main points are summarized below:

1. Ecological communities are formed by a web of interacting populations, food webs or food chains, with plants forming the base production level and herbivores and carnivores forming a trophic pyramid above. The length of food chains is limited by the requirements of conservation and the loss of energy in transferring biomass between trophic levels.
2. Community network models were constructed by considering the density (or biomass) of each species as a state variable with feedback between members of the same species represented by self-loops and between different species by competitive, cooperative, or trophic interaction loops.
3. The overall feedback structure of a community was defined by the sum of the feedback at each level of organization, or by what is called loop analysis; i.e., feedback at level 1 is the sum of all self-loops (ΣS_{ij}), at level 2 it is the sum of

all conjunct two-species interactions ($\Sigma C_{ij} C_{ji}$) minus the sum of disjunct loop products involving two species ($\Sigma S_{ji} S_{jj}$), and so on. The general expression for feedback at level k is given by

$$F_k = \sum_{m=1}^{k} (-1)^{m+1} L(m,k),$$

where m is the number of loops involved in each feedback term (Note 6.3).

4. A community is defined as being stable in the vicinity of its equilibrium position if feedback at all levels of organization is negative, with the proviso that feedback at levels 3 or higher must be weaker than the product of feedback at lower levels; e.g.,

$$F_1 F_2 + F_3 > 0.$$

When negative feedback at the higher levels is strong, relative to that at lower levels, the effect of time delays in the long loops may give rise to oscillatory instability.

Loop analysis was used to evaluate the stability properties of ecological communities, from which the following generalizations emerged:

5. Competitive interactions between species create unstable communities unless they are dominated by negative feedback between members of the same species (self-loops). In addition, as more competing species are added to the community, the competitive interactions must become proportionally weaker if community stability is to be maintained, because the constraints on stability become more restrictive at higher levels of organization; i.e., stability decreases as more species are added unless the competitive interactions weaken through evolutionary adjustments.

6. Trophic (predator–prey) interactions are inherently stabilizing and may, under certain conditions, stabilize otherwise unstable interactions between competing species; i.e., if the predators are very efficient and the intensity of competition between their prey is directly related to their density, or if the predators preferentially crop the more abundant prey species. From this we concluded that predation can increase the diversity and stability of ecological communities.

7. Given enough time in a consistent environment, species can evolve behavioral adaptations that modify their competitive, cooperative, and trophic interactions and, thereby, a stable community may be created. Hence, complex and diverse communities have a greater probability of evolving in benign and stable environments.

8. Community stability is frequently disrupted by severe environmental disturbances and this is usually followed by a series of successional communities that slowly evolve towards a climax association. However, continuous disturbances may prevent the attainment of this climax because the community is kept in its early successional stages. The climax community is adapted to particular climatic

and edaphic conditions so that we often find a spatial mosaic of different communities separated by distinct topographic and edaphic boundaries, and each of these communities may be in different stages of succession.

9. Herbivores play an important role in community succession by removing certain individuals and species, and thereby, hastening the rate of succession. In addition, herbivores often feed on the unproductive, or unhealthy, members of the community. In this way they act to recycle nutrients to the more productive components, and to increase the overall productivity and vigor of the community. From this point of view, trophic interactions can be considered mutualistic at the community level as both prey and predator benefit from the association.

Exercises

6.1. In a forest community, fungal pathogens, which slowly debilitate and eventually kill trees, are spread more readily in dense stands However, these fungus-infected trees are usually attacked by bark beetles before they die from the fungus infection. The beetles kill these weak trees and then breed within them. Assuming that stand density is self-limiting, draw the feedback diagram for this community and evaluate its stability.

6.2. Some bark beetles have evolved a mechanism for transporting fungi and inoculating them into the tree as they attack it. These fungi then aid the beetle in killing the tree. Has this evolutionary trend contributed to the stability of the fungus–beetle–tree interaction, and if not, what conditions are necessary for a stable interaction?

In the case of the Dutch elm disease, beetles inoculate fungi into trees during maturation feeding; the fungi then cause blockage of the tree's food transport system and severely weaken its defenses so that the beetles can easily gain entrance and kill the tree. In other words, we have a very strong mutualistic interaction between beetle and fungus. In the light of your previous answer, what can you conclude about the stability of this system?

6.3. Hares feed on vegetation and are themselves fed upon by lynx. However, the intensity of lynx predation is reduced by heavy vegetation because the hares are harder to find. Assuming that vegetation is self-limiting, evaluate the stability of this community. What conditions are necessary for oscillatory instability, which may give rise to observed population cycles? What evolutionary trends would stabilize this system?

6.4. Grass, a self-limited resource, forms the food supply for an antelope herd. A plentiful supply of grass increases the health and vigor of the herd and its resistance to a pathogenic microorganism. When the general health of the herd is low, epidemics of the pathogen kill many weakened antelope. Naturally, healthy herds produce more surviving young and, thereby, increase the size of the herd. Evaluate the stability of this system.

Notes

6.1. The question of complexity versus stability of mathematical models of eco-
    logical systems has been addressed exhaustively by Robert May in his book
    Stability and Complexity in Model Ecosystems, published by Princeton
    University Press, New Jersey, 1974. The overriding conclusion of most math-
    ematical exercises is that complex systems are never more stable, and are usu-
    ally less stable, than simple systems. However, when more realistic biological
    features are incorporated into the models, the contrary becomes true again,
    as exemplified, for instance, in the paper by M. Rejmánek, P. Kindlmann and
    J. Lepš: Increase of stability with connectance in model competition communities,
    published in the J. Theor. Biol. (vol. 101, pp. 649–656, 1983).
6.2. Evaluating the qualitative stability of complex systems is briefly covered in
    May's book (Note 6.1). However, Richard Levins provides a much more com-
    plete treatment in his contribution to the book *Ecology and Evolution of
    Communities*, which was edited by M. L. Cody and J. M. Diamond and pub-
    lished by The Belknap Press of Harvard University, Cambridge, Massachusetts,
    in 1975. Levin's contribution is suggested for those who wish to explore the
    mathematical details of loop analysis. Levins' paper in the *Annals of the New
    York Academy of Sciences* (vol. 231, p. 123, 1974) also discusses loop analysis
    and its applications in biology.
6.3. The general expression of feedback at the kth level of organization is

$$F_k = \Sigma(-1)^{m+1} L(m,k)$$

where m is the number of loops involved in the computation of each feedback
term. The first expression in this equation, $(-1)^{m-1}$, adjusts the sign so that it
is always negative when the loops in a particular product are all negative.
Thus, when $k = 1$, feedback at level 1 is

$$F_1 = \Sigma(-1)^{1-1} S_{ij} = \Sigma\, S_{ii}$$

because only one loop is involved in each addition. However, at level 2 we
have single loops involving two species, $C_{ij}\, C_{ji}$, and combinations of two self-
loops. Thus,

$$F_2 = \Sigma\, (-1)^{1-1} C_{ij} C_{ji} + \Sigma\, (-1)^{2+1} S_{ii} S_{jj}$$
$$= \Sigma\, C_{ij} C_{ji} - \Sigma\, S_{ii} S_{jj}$$

At the third level we will have single loops involving three species whose
signs are positive; disjunct loops composed of a self-loop and a two-species
connection, whose sign is negative; and disjunct loops composed of three self-
loops, whose sign is $(-1)^{3-1} = +1$.

$$F_3 = \Sigma\ C_{ij}C_{jk}C_{ki} - \Sigma\ S_{ii}C_{jk}C_{kj} + \Sigma\ S_{ii}S_{jj}S_{kk}$$

and so on for higher levels. For further details the reader is referred to Levins' writings mentioned in Note 6.2.

6.4. The conclusion that environmental stability permits communities to become more diverse or complex has a basis in information theory. The rate of information flow through an information system is reduced if the channel is noisy (has a lot of static) because the signals cannot be so finely divided. For instance, try communicating very rapidly over the airwaves when a lot of static is present. Similarly, in a noisy environment (i.e., a variable one) evolution cannot proceed at as fast a pace as in a quiet (consistent) one and, therefore, fewer species will be present. The fundamental theorem of information theory is expressed by

$$V = A \times \log_e(1 + B/n),$$

where V is the rate of evolution (analogous to information flow) which is assumed proportional to the number of species in the community, n is a measure of environmental variation (noise), and A and B are constants, perhaps related to the mutation rate and the diversity of the base resources, respectively. John MacArthur, in his contribution to the book *Ecology and Evolution of Communities* (see Note 6.2 for the complete reference), obtained remarkably good fits to this simple model with data from bird, mammal, and gastropod diversity gradients along latitudinal transects, where the measure of environmental variation (noise) was the difference in mean winter–summer temperatures.

6.5. One of the most often cited experiments on predator removal is that published by R. T. Paine in 1966 in the *American Naturalist* (vol. 100, p. 65). Paine removed the predatory starfish, *Pinaster*, from an area of seashore and found that the community diminished from 15 to 8 species within 2 years. Support for this view can also be found in the works of D. J. Hall, W. E. Cooper, and E. E. Werner in *Limnology and Oceanography* (vol. 15, p. 839, 1970) and in J. H. Connell's contribution to the book *Dynamics of Populations*, edited by P. J. den Boer and G. R. Gradwell, and published by the Centre for Agricultural Publishing and Documentation (Wageningen, Netherlands, 1971). Connell further summarizes this empirical evidence for the role of predation in community diversity in his contribution to the book *Ecology and Evolution of Communities* (see Note 6.2 for the reference).

There have also been a number of mathematical analyses of the effect of predation on the stability and diversity of ecological communities. These are summarized by R. M. May in his book *Stability and Complexity in Model Ecosystems* (Note 6.1), and by M. P. Hassell in his book *Dynamics of Arthropod Predator–Prey Systems* (Princeton University Press, New Jersey, 1978).

6.6. The competitive interaction between ponderosa and Jeffrey pines, and their hybridization in transitional areas, was described by J. R. Haller in *University of California Publications in Botany* (vol. 34, p. 123). He showed that the two species coexist in the transitional zone because both are tolerant of the physical environment, that hybrids cannot compete with their parents, and that the intermixed serpentine and basaltic soils provide each species with competitive advantages.

6.7. A book *Theory and Practice of Mountain Pine Beetle Management in Lodgepole Pine Forests*, edited by A. A. Berryman, G. D. Amman, R. W. Stark, and D. L. Kibbee and published by the Forest, Wildlife, and Range Experiment Station of the University of Idaho, Moscow (1978), provides detailed information on this forest–insect interaction. A contribution by R. L. Mahoney in this book shows that outbreaks of the beetle tend to start in stands where the growth rate is in decline.

6.8. The proposition that herbivorous insects act as regulators of forest productivity and nutrient cycling and, thereby, perform a vital function in the dynamics of ecological communities, was examined by W. J. Mattson and N. D. Addy in *Science* (vol. 190, p. 515, 1975). They concluded that insect grazers function like feedback regulators of primary productivity, ensuring consistent and optimal output of plant production over the long term on a given site. This conclusion was based on the observation that the activity of herbivores was often inversely related to the vigor and productivity of the plant community. As a result of this interaction, they suggested that nutrients are cycled from the nonproductive components of the system to the more vigorous, productive elements and, because of this, insect–plant relationships may be considered mutualistic in the long-term sense. This line of reasoning was also taken by R. M. Peterman in his contribution to the book mentioned in Note 6.7. He argued that mountain pine beetle populations and forest fires usually destroyed lodgepole pine stands at a time, which maximizes their long-term fitness and productivity.

Epilogue
The Human Dilemma

Several million years ago a group of apelike animals emerged from the East African savanna with new attributes to test in the arena of evolution - a grasping tool-wielding hand, and the glimmerings of intelligence and cooperative social organization. So powerful were these new adaptations that the species we know as *Homo sapiens* spread rapidly from its African genesis to all corners of the planet Earth. No other single species has been so successful in the evolutionary struggle for dominance, even the fierce Pleistocene predators yielding in the face of cooperative intelligence and hand-wielded weapons. Today *Homo sapiens* stands at the pinnacle of his power - proud, indomitable, and confident in his ability to meet the next evolutionary challenge.

And yet a threat looms on his horizon.

Not a threat from predators or competitors, but from man's own cooperative abilities to overwhelm the negative feedback (diseases, predators, competitors, food shortage, etc.) acting on his populations. As we know, the dominance of cooperative interactions creates an unstable positive feedback loop (Chapter 3), an instability, which is reflected in the alarming growth of the human population of more than six billion individuals. However, we also know that negative feedback in the form of competitive interactions must eventually dominate a population inhabiting a finite environment. It is the threat of fulfilling the Malthusian prophecy, which looms over the future of mankind.

We cannot refute the basic Malthusian premise that the earth and its resources are finite, or the resulting deduction that the human population cannot grow indefinitely. There must be some equilibrium density, or carrying capacity for the planet, which will determine the population that can be sustainable indefinitely by the earth's resources. However, although *Homo* may have the intellect to arrive at this conclusion, he has yet to calculate a value for this carrying capacity. In this respect he is no better off than other less intelligent species, which are subjected to competitive struggle and subsistence economy whenever their populations approach or exceed carrying capacity.

Because there is no firm estimate of the human carrying capacity, it is impossible to predict the future behavior of the population with any degree of accuracy. Lacking facts, predictions have come from the mouths of prophets and soothsayers. The pessimistic believe that we have already surpassed this unknown equilibrium density and that we are doomed to a future of misery, vice, and destruction as the price of overshoot and overexploitation. Even the more optimistic realize that the population explosion has caused severe impacts on its environment, and that the time delays introduced into the negative feedback loops may create cycles of growth and collapse, prosperity and misery. On the other hand, the most optimistic believe that human intelligence, ingenuity, and cooperation can continue to raise the carrying capacity of the globe and correct the negative impacts on the environment before they are fed back to future generations. Indeed, there is some precedent for this optimistic view because the dismal prophecies of Thomas Malthus, almost two hundred years ago, failed to foresee the unprecedented advances in agricultural technology, which have greatly increased the carrying capacity of the earth. There is a danger, however, in relying on technology to continue to expand the carrying capacity ahead of the growing population. Agricultural technology is based, in large part, on a finite and dwindling supply of energy - the Malthusian premise remains. Moreover, even if technology can harness an infinite energy source, such as the sun, negative feedback must eventually dominate as more and more agricultural land is used for living space or for solar conversion devices. The question then is not whether the human population can continue to grow *ad infinitum*, but rather *when* and *how* it will be brought under control, and whether it will suffer the drastic consequences of overshooting its carrying capacity.

Aside from the serious problem of the expanding human population, which can only be solved by social, cultural, or political adjustments, there remains the question of how best to manage our renewable natural resources for the benefit of present and future generations. It seems to us that many of our ecological problems have roots in our humanitarian philosophies. Early humanitarian concepts, which form the cornerstones of Western civilizations, center on the rights of individuals to compete with equal opportunity for food, material wealth, and social well-being. Although it is difficult to argue with this principle, it has one fundamental flaw. The egalitarian ideals were formed at a time when opportunity and resources seemed inexhaustible and the human population was relatively small. Because of this they did not adequately consider the rights of unborn generations. Thus, in the name of human rights and equal opportunity we have plundered and squandered the earth's resources without concern for the rights, needs and opportunities of future peoples. In this era of *exploitation*, supplies of fuels and minerals have been severely depleted and populations of animals harvested to extinction, or near extinction. This has led us into the ecological crisis of today, and the confrontation between exploiter and conservationist.

Forged out of the ecological crisis a new ethic has emerged which recognizes the rights of generations yet unborn. This is the ethic of *conservation*, the imperative of the predator to leave resources to nourish its offspring (Chapter 4). Perhaps it is incorrect to claim this as a new ethic for surely early human populations conserved their food supply in a similar way to other predators. Certainly it was not the American Indian who overexploited the bison herds and salmon runs! But then again early man was not the efficient hunter he is today, and conservation was not such a critical imperative.

The ethic of conservation, recognizing the rights of future generations and conserving resources for their use, has become the guiding philosophy of the manager of natural renewable resources. From this philosophy has sprung the concept of sustainable yield, or setting harvest policies that permit utilization of the resources in perpetuity. Although sustainable yield is not a new idea - having been practiced by central European foresters for centuries - the maintenance of an effective sustainable yield policy is no simple matter, for the manager is faced with a complex array of biological, social, and economic problems. The manager needs to have a clear understanding of the interaction structure of the population system he is managing and the consequences of his decisions. For instance, he has to consider the consequences of time delays, not only within the population system itself, but also within the management cycle, for it takes time to implement management policies and by then things may have changed. He should be aware that time delays can give rise to population cycles, and that even if cycles are not apparent they can be created if the environment is altered by management practices (Chapter 2). The manager must be wary of thresholds in the system created by cooperative interactions, particularly extinction thresholds, from which there is no return (Chapters 3 and 4). Harvesting policies should be formulated to minimize the risks of crossing these unstable equilibria and precipitating undesirable population behavior. On the other hand, thresholds can sometimes be used to advantage, as in the biological control of pest species. The manager also needs to have a feeling for the larger ecosystem, of which the population he is managing is part, for policies implemented on one species and at one place are likely to influence other species in different places (Chapters 5 and 6). As if these ecological problems are not enough, the resource manager is often confronted with even more difficult economic and social problems, which may conflict with his conservative ethics. Here we reach the crux of the management dilemma.

The success of a sustainable yield policy requires that the supply of a renewable resource to its human consumer be regulated in such a manner that it is maintained indefinitely into the future. Although this supply rate can be raised or lowered by cultural practices that alter the favorability of the environment for the species being managed (Chapter 3), it must be regulated independently of demands by the consumer if the sustainable yield policy is to be successful. Society, however, views the problem from the side of the consumer rather than the resource. Thus, socialists demand more resources to raise the standard of living of the workers, while capitalists demand more to increase profits for their shareholders. If the resource manager regulates the supply, S, and social pressures create the demand for resources by each individual in the population, d, then the management dilemma is captured in the supply/demand ratio

$$S \: / \: dN,$$

where N is the size of the consumer population, and dN is the total demand for resources made by the population. As long as the supply of resources from a sustainable yield policy exceeds the total demand of consumers (i.e., $S/dN > 1$) the manager only has to worry about his biological problems. However, as soon as the demand exceeds the supply (i.e., $S/dN < 1$) he finds himself in a very difficult position: He can either abandon his sustainable yield policy and increase the supply, compromising his responsibilities to the future and risking extinction of the resource, or he can insist on maintaining a constant supply and incur the wrath of his fellows whether they be socialists or capitalists. Because the resource manager is often a public servant and, therefore, subject to political pressures, it may be impossible for him to resist the demands from both left and right wing political fractions, and he may be forced to abandon his sustainable yield policy, regardless of his ethical standards. The manager, it seems, is often caught "between the devil and the deep blue sea."

It appears that we always have to return to the central ethical problem: Do future unborn generations have a right to a share of the earth's resources? Even the most adamant conservationist is burdened with the same question whenever he drives his automobile to the supermarket and, thereby, burns the fuel which can be used to grow food for his descendants. If we choose to accept this moral standard, then conservation and sustainable yield become the maxima of natural resource management. Under these maxima, an ecological solution is only possible by reducing the consumer demand to meet the sustainable yield supply. This can be done either by reducing the demand of each individual, d, which is the same as lowering the standard of living, or by stabilizing or reducing the population size, N, so that

$$dN = S.$$

Because population size is not easily or quickly changed, the only viable alternative seems to be through the standard of living. The usual economic solution to this problem is to allow the price of the resource to rise with the demand for it. In other words, as the resource becomes scarce, relative to the consumer demand for it, the price rises so that more individuals are forced out of the market. In this way the population of consumers, N, is reduced to those who can afford to pay the price. The economic equilibrium then becomes

$$dN / p = S,$$

or

$$p = dN / S,$$

where p is a pricing coefficient. The economic solution may be socially acceptable for nonessential commodities such as automobiles and washing machines. However, most renewable resources are used as food, clothing, or housing, and so this solution violates our humanitarian standards because the poor are deprived of the necessities of life.

It seems to us that many of our present-day ecological problems are rooted in economic attitudes. For example, many economists reject the concept of sustainable yield, arguing that such policies neglect the industrial costs of harvesting and marketing the resource. Rather than maximizing the sustainable yield, they say, the manager should attempt to maximize the net income derived from the resource. Unfortunately, this policy can lead, under certain conditions, to the resource being harvested to extinction (see for example C. W. Clark, *Science*, vol. 181, p. 630, 1973). Once again we come up against the ethical question of descendant rights.

Contemporary economic theory has evolved, to a large extent, from the ideas of David Ricardo, a vigorous proponent of economic growth. Ranged against him, in what was one of the friendliest controversies in the history of intellectual pursuit, was another economist and population theorist, Thomas Malthus. That Ricardo emerged from this debate as the overwhelming victor is one of the ironies of our times but not surprising, knowing man's innate optimism and greed. John Maynard Keynes laments in his biography of Malthus:

> One cannot rise from a perusal of this correspondence [between Ricardo and Malthus] without a feeling that the almost total obliteration of Malthus's line of approach and the complete domination of Ricardo's for a period of a hundred years has been a disaster to the progress of economics. ... If only Malthus, instead of Ricardo, had been the parent stem from which nineteenth-century economics proceeded, what a much wiser and richer place the world would be today!

(J. M. Keynes, *Essays in Biography*, New Edition, London, 1951)

One can hardly resist from adding to Keynes's lament that, if only Malthus's basic concepts of population, if not his methods, had played a central role in political–economic thinking and population planning, what a much more pleasant and bountiful planet we could bequeath to our (limited) offspring.

Answers to Exercises

Chapter 2

2.1. (A) $R = 0.08$; (B) 108, 117, 126, 136, 147; (C) It assumes growth is unlimited; (D) As $R = 0$, it will remain at the same density.

2.2. (A) Negative feedback; (B) $R = 1.5$; (C) $R_m = R/(1 - N/K) = 1.5/(1 - 200/2000)$ $= 1.67$; (D) $s = R_m/K = 0.00083$; (E) 500, 1126, 1948, 2033, 1977, 2015, 1990, 2007, 1995, 2003; (F) Damped-stable approach to equilibrium because $y/x = 10/15 = 0.6$ and $R_m T = 1.5$.

2.3. (A) 990, 998, 1000, 1000, etc.; $y = -2$, $y/x = -0.2$; $R_m T - 1 = -0.2$, asymptotic stability; (B) 9990, 9998, 10000, 10000, etc.; $y/x = -0.2$, $R_m T - 1 = -0.2$, asymptotic stability; (C) Similar result: (D) 990, 1008, 994, 1005, 996, 1003; $y/x = 0.8$, $R_m T - 1 = 0.8$, damped stable oscillations; (E) 990, 1018, 967, 1056, 890, 1164; $y/x = 1.8$, $R_m T - 1 = 1.8$, unstable; (F) 990, 990, 998, 1006, 1008, 1003; $y/x = 0.8$, $R_m T - 1 = 0.6$, damped stable oscillations; (G) 990, 990, 990, 998, 1006, 1014, 1016, 1011; $y/x = 1.6$, $R_m T - 1 = 1.4$, unstable.

Chapter 3

3.1. (A) 17.4, −0.57, 1.0, −0.48, 0.64, −0.25, 0.25, −0.25, 0.2; (B) Damped stable oscillations because $y/x \approx 0.6$; $K \approx 175$, $s \approx 0.01$, $sK \approx 1.8$; (C) No evidence for cooperative interactions; $T \approx 1$; no long time delays because environmental feedback is minimized by replacing food at the start of each generation.

3.2. (A) 0.6, −0.69, 3.0, −0.93, 1.67, 3.25, −0.79, 3.29, −0.47, 1.31; (B) Unstable because $y/x \approx 1.3$; $K \approx 250$, $s \approx 0.01$, $sK \approx 2.5$; (C) Seems to be globally stable under most conditions because oscillations do not continue to increase in amplitude indefinitely. However, there is evidence for an extinction threshold at a population density between 10 and 20; (D) Cooperative low-density interactions are evident, $T \approx 1$.

3.3. **(A)** 2.03, 3.18, 0.05, −0.38, −0.78, −0.26, 1.04; **(B)** $T > 1$ because a cyclic trajectory is evident. A plot of R on N_{t-2} yields an approximately single-line relationship, thus, $T \approx 2$.

3.4. **(A)** The oak environment is more favorable, providing a higher equilibrium density, K, and a larger value for sK, hence, the more vigorous oscillations; **(B)** $T \approx 1$ because there is little tendency towards population cycles.

3.5. $K \approx 20$ in unthinned woods, $K \approx 47$ in thinned woods; an appropriate reproduction plane and equilibrium line can be drawn with stand density as the environmental favorability axis.

3.6. If we start at the beginning of the second cycle (the year 1919) we can find that hare numbers, $H_0 \approx 20$, and lynx numbers, $L_0 \approx 2$. In the next year the lynx population increased to 5, giving a net reproduction of 3 lynx. We plot this first population vector as a horizontal arrow from 2 to 5 lynx opposite a hare density of 20. The environmental change vector is then calculated as $H_1 - H_0 = 38 - 20 = 18$, and plotted as a broken line (Figure A). The next lynx change from 5 to 15 is plotted from this point, and so on. We can place the equilibrium line approximately knowing that the lynx population increases to its left and decreases to its right (Figure A).

3.7. Because there is no evidence for a drastic *permanent* change in the favorability of the salmon's environment, we should suspect that the system has a complex W-shaped equilibrium line. The salmon population cycles around its upper equilibrium level, indicating the action of delayed feedback operating through the environment (gene pool?). However, cycles are less evident in the domain of the lower equilibrium, suggestive of rapid (non-delayed) density-dependent

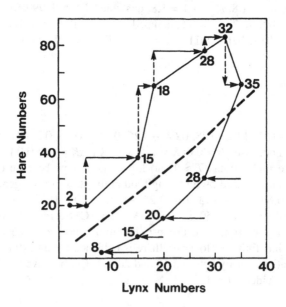

Fig. A

responses. Thus, the critical density, N_c, where the population affects the qualities of its environment, probably lies somewhere between the two equilibria.

Chapter 4

4.1. **(A)** The repressive effect of each individual on the reproduction and survival of its cohorts, s, and on the other species, c; **(B)** (i) A and B coexist; (ii) B replaces A; (iii) A or B wins depending on the starting densities. Populations coexist when $c_a < s_a$ and $c_b < s_b$; **(C)** A replaces B because $R_{ma}/R_{mb} = s_a/s_b$; **(D)** Details of equilibrium and extinction behaviors. You may also obtain negative population densities, an unreasonable feature of the linear models.

4.2. See Figure 4.12 (page 113).

4.3. Mobility (finding resources quickly), high maximum rate of increase (advantage of numbers), and life in variable or temporary habitats.

4.4. **(A)** The vulnerability of the prey to attack and the efficiency of the predator in converting prey into predator offspring; **(B)** Equilibrium at $A = 500$, $B = 100$; damped-stable cycles; for instance, when $A_0 = A^* - 100$, $B_0 = B^* - 40$ we get the following dynamics: $A = 640, 620, 433, 398, 588, 608, 447, 408, 568, 596, 459, 417$; $B = 75, 106, 121, 73, 79, 105, 119, 79, 81, 104, 117, 83$; **(C)** $A = 286$, $B = 143$; unstable cycles of increasing amplitude; for example, when $A_0 = A^* - 4$, $B_0 = B^* + 7$, we obtain $A = 267, 272, 308, 316, 269, 238, 290, 360, 317, 207, 209, 381, 484$; $B = 145, 132, 136, 152, 158, 130, 118, 140, 171, 157, 76, 97, 144$; **(D** $A = 444$, $B = 222$; stable cycles; for example, when $A_0 = A^* - 56$, $B_0 = B^* + 58$, we obtain $A = 300, 350, 651, 676, 403, 300, 476, 620, 456, 327, 409, 597, 520$; $B = 246, 88, 132, 210, 290, 162, 149, 204, 274, 219, 144, 187, 267$.

4.5. The predator reproduction plane is similar to that in Figure 4.17C and the prey's is like Figure 4.18A. The interaction will produce stable limit cycles.

4.6. **(A)** About 178; **(B)** Prey equilibrium density to about 147; **(C)** New prey equilibrium at about 12; **(D)** At least 52.

Chapter 6

6.1. (See Figure B).

$$F_1 = - S_{tt}$$
$$F_2 = - C_{tb} C_{bt} - C_{tf} C_{ft}$$
$$F_3 = - C_{tf} C_{fb} C_{bt}$$
$$F_1 F_2 + F_3 = S_{tt} C_{tb} C_{bt} + S_{tt} C_{tf} C_{ft} - C_{tf} C_{fb} C_{bt} > 0$$

The community is stable and oscillatory instability seems unlikely under most conditions.

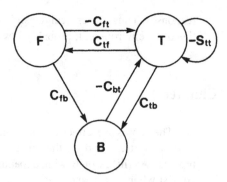

Fig. B

6.2. Add another positive interaction C_{bf} for the beetle helping the fungus to the community above.

$$F_1 = - S_{tt}$$
$$F_2 = - C_{tb} C_{bt} - C_{tf} C_{ft} + C_{fb} C_{bf}$$
$$F_3 = - C_{tf} C_{fb} C_{bf} - C_{tb} C_{bf} + Stt \, C_{fb} C_{bf}$$

The community is less stable because positive terms have been added to level-2 and -3 feedback. The system will be stable as long as feedback between beetle and fungus is not too strong. In the case of the Dutch elm disease the community is likely to be unstable because the feedback between beetle and fungus is very strong. This conclusion is borne out by the facts as the disease has all but eliminated American elms from the eastern and central USA and is currently sweeping through European elm forests.

6.3. (See Figure C).

$$F_1 = - S_{vv}$$
$$F_2 = - C_{vh} C_{hv} - C_{hl} C_{lh}$$
$$F_3 = - C_{vl} C_{lh} C_{hv} - S_{vv} C_{hl} C_{lh}$$
$$F_1 F_2 + F_3 = S_{vv} C_{vh} C_{hv} - C_{vl} C_{lh} C_{hv} > 0$$

Factoring out C_{hv} we find that the system will be stable

$$C_{vl} C_{lh} > S_{vv} C_{vh}$$

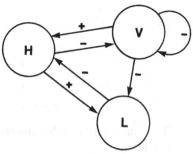

Fig. C

Evolutionary trends towards lower vulnerability of hares to predation (decreases C_{lh}), increased efficiency of lynx finding hares in dense cover (decreases C_{vl}), increased efficiency of hares in utilizing the vegetation (increases C_{vh}), or more powerful self-limitation of vegetation will all increase stability.

6.4. (See Figure D).

$$F_1 = -S_{gg}$$

$$F_2 = -C_{da} C_{ad}$$

$$F_3 = -C_{gh} C_{ha} C_{ag} - S_{gg} C_{da} C_{ad}$$

$$F_1 F_2 + F_3 = C_{gh} C_{ha} C_{ag} = \text{oscillatory instability}$$

G = grass, H = health of the herd, A = size of the herd, D = disease.

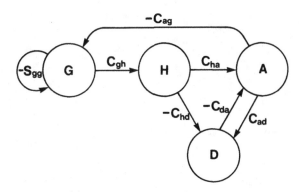

Fig. D

Name Index

Subject Index